Phase Portraits of Control Dynamical Systems

Mathematics and Its Applications (*Soviet Series*)

Volume 63

Phase Portraits of Control Dynamical Systems

by

Anatoliy G. Butkovskiy

*Laboratory for Distributed Parameter Systems,
Institute of Control Sciences, Moscow, U.S.S.R.*

SPRINGER SCIENCE+BUSINESS MEDIA, B.V.

Library of Congress Cataloging-in-Publication Data

Butkovskiĭ, A. G. (Anatoliĭ Grigor´evich)
    [Fazovye portrety upravlíaemykh dinamicheskikh sistem. English]
    Phase portraits of control dynamical systems / by Anatoliy G.
Butkovskiy.
        p.   cm. -- (Mathematics and its applications (Soviet series) ;
v. 63)
    Translation of: Fazovye portrety upravlíaemykh dinamicheskikh
sistem.
    Includes bibliographical references and index.
    ISBN 978-94-010-5437-9   ISBN 978-94-011-3258-9 (eBook)
    DOI 10.1007/978-94-011-3258-9

    1. Differentiable dynamical systems. 2. Control theory.
I. Title. II. Series: Mathematics and its applications (Kluwer
Academic Publishers). Soviet series ; 63.
QA614.8.B88   1991
515'.35--dc20   90-22693

ISBN 978-94-010-5437-9

SERIES EDITOR'S PREFACE

'Et moi, ..., si j'avait su comment en revenir, je n'y serais point allé.'

Jules Verne

The series is divergent; therefore we may be able to do something with it.

O. Heaviside

One service mathematics has rendered the human race. It has put common sense back where it belongs, on the topmost shelf next to the dusty canister labelled 'discarded nonsense'.

Eric T. Bell

Mathematics is a tool for thought. A highly necessary tool in a world where both feedback and non-linearities abound. Similarly, all kinds of parts of mathematics serve as tools for other parts and for other sciences.

Applying a simple rewriting rule to the quote on the right above one finds such statements as: 'One service topology has rendered mathematical physics ...'; 'One service logic has rendered computer science ...'; 'One service category theory has rendered mathematics ...'. All arguably true. And all statements obtainable this way form part of the raison d'être of this series.

This series, *Mathematics and Its Applications*, started in 1977. Now that over one hundred volumes have appeared it seems opportune to reexamine its scope. At the time I wrote

"Growing specialization and diversification have brought a host of monographs and textbooks on increasingly specialized topics. However, the 'tree' of knowledge of mathematics and related fields does not grow only by putting forth new branches. It also happens, quite often in fact, that branches which were thought to be completely disparate are suddenly seen to be related. Further, the kind and level of sophistication of mathematics applied in various sciences has changed drastically in recent years: measure theory is used (non-trivially) in regional and theoretical economics; algebraic geometry interacts with physics; the Minkowsky lemma, coding theory and the structure of water meet one another in packing and covering theory; quantum fields, crystal defects and mathematical programming profit from homotopy theory; Lie algebras are relevant to filtering; and prediction and electrical engineering can use Stein spaces. And in addition to this there are such new emerging subdisciplines as 'experimental mathematics', 'CFD', 'completely integrable systems', 'chaos, synergetics and large-scale order', which are almost impossible to fit into the existing classification schemes. They draw upon widely different sections of mathematics."

By and large, all this still applies today. It is still true that at first sight mathematics seems rather fragmented and that to find, see, and exploit the deeper underlying interrelations more effort is needed and so are books that can help mathematicians and scientists do so. Accordingly MIA will continue to try to make such books available.

If anything, the description I gave in 1977 is now an understatement. To the examples of interaction areas one should add string theory where Riemann surfaces, algebraic geometry, modular functions, knots, quantum field theory, Kac-Moody algebras, monstrous moonshine (and more) all come together. And to the examples of things which can be usefully applied let me add the topic 'finite geometry'; a combination of words which sounds like it might not even exist, let alone be applicable. And yet it is being applied: to statistics via designs, to radar/sonar detection arrays (via finite projective planes), and to bus connections of VLSI chips (via difference sets). There seems to be no part of (so-called pure) mathematics that is not in immediate danger of being applied. And, accordingly, the applied mathematician needs to be aware of much more. Besides analysis and numerics, the traditional workhorses, he may need all kinds of combinatorics, algebra, probability, and so on.

In addition, the applied scientist needs to cope increasingly with the nonlinear world and the extra mathematical sophistication that this requires. For that is where the rewards are. Linear models are honest and a bit sad and depressing: proportional efforts and results. It is in the non-linear world that infinitesimal inputs may result in macroscopic outputs (or vice versa). To appreciate what I am hinting at: if electronics were linear we would have no fun with transistors and computers; we would have no TV; in fact you would not be reading these lines.

There is also no safety in ignoring such outlandish things as nonstandard analysis, superspace and anticommuting integration, p-adic and ultrametric space. All three have applications in both electrical engineering and physics. Once, complex numbers were equally outlandish, but they frequently proved the shortest path between 'real' results. Similarly, the first two topics named have already provided a number of 'wormhole' paths. There is no telling where all this is leading - fortunately.

Thus the original scope of the series, which for various (sound) reasons now comprises five subseries: white (Japan), yellow (China), red (USSR), blue (Eastern Europe), and green (everything else), still applies. It has been enlarged a bit to include books treating of the tools from one subdiscipline which are used in others. Thus the series still aims at books dealing with:

- a central concept which plays an important role in several different mathematical and/or scientific specialization areas;
- new applications of the results and ideas from one area of scientific endeavour into another;
- influences which the results, problems and concepts of one field of enquiry have, and have had, on the development of another.

A picture is worth a thousand words. Indeed in the present case, I would be hard put to describe in only 1000 words the insight and information contained in these phase portraits of control systems by Butkovskii.

Making a phase portrait of a single dynamical system (an O.D.E.) is already a nontrivial task. In the case of systems with controls one essentially is faced with the challenge of portraiting a (multi dimensionally) infinite family of such dynamical systems. The author has found an illustrative way of doing this by focussing (which is natural) on the 'boundaries' of integral funnels of trajectories of the differential inclusions involved.

The result is a unique and uniquely instructive book, which, even though this is a straightforward translation of a Russian edition that appeared a non-negligible number of years ago, has lost none of its appeal and uniqueness.

The author has on occasion adopted a somewhat less formal approach to present (and discover) his insights and as a result the work also contains not a few challenges to the more rigorously mathematically inclined in making these discoveries of the author mathematically absolutely precise.

I consider this one an invaluable book (and so did my advisors) and am happy to present it in this series to the (very critical) world of mathematicians and applications at large.

The shortest path between two truths in the real domain passes through the complex domain.

J. Hadamard

La physique ne nous donne pas seulement l'occasion de résoudre des problèmes ... elle nous fait pressentir la solution.

H. Poincaré

Never lend books, for no one ever returns them; the only books I have in my library are books that other folk have lent me.

Anatole France

The function of an expert is not to be more right than other people, but to be wrong for more sophisticated reasons.

David Butler

Bussum, October 1990

Michiel Hazewinkel

CONTENTS

Preface to the Soviet Edition

The present book is the outcome of efforts of giving vivid geometrical representation to control dynamical systems (CDSs) which occur in the theory and practice in several scientific and technical disciplines. The fundamental question with which the author was concerned was how to make use of the phase space (state space, and, in particular, the phase plane of CDS) in giving as complete as possible a description of such systems.

Unlike the qualitative theory of usale dynamical systems (UDSs) in which one is concerned with differential equations (without control parameters), in the discussion of a CDS one has to deal with differential inclusions, adequately representing control dynamical systems. In this connection the author was faced with the following question: what is to be understood by the phase portrait of a differential inclusion or, in other words, what is the phase portrait of a CDS ? The author supposes that he has found a way of attacking this problem in the case of a two-dimensional CDS by means of boundaries of integral and trajectory funnels of the differential inclusion. The concepts of boundaries of integral and trajectory funnels play a significant role in the general case too. But here the general phase portraits of CDSs are more complex and more varied. In this case only a general plan of investigation is suggested, though some general and particular notion and facts are established.

Regarding the style of the present book, it should be remarked that the author has tried to find the least formal and, as far as possible, vivid way of presenting the subject matter, sometimes even at the cost of rigour. The author fully realizes that some of the ideas and results presented possibly need to be made precise and subjected to a detailed analysis. In such cases the author has taken recourse to such phrases as "in general" or "generally speaking" thereby underlining that in such cases rigour and precision possibly require reformulation and modification in the statement in question. Such a presentation of the subject matter seems to be useful when the theory of phase portrait of CDS is at the present stage of development.

It is also necessary to remark that since there is already a vast literature available on the qualitative theory of dynamical control systems, the references given at the end of the book do not claim to be exhaustive or even fairly complete.

In the preparation of the book a positive role was played by the discussion the author had with his colleagues in private as well as in public meetings. The author expresses his

1

2

gratitude to them. Special thanks are due to M.A. Aizerman, M.A. Krasnosel'skii, V.F. Krotov, V.P. Maslov, A.F. Fillipov, F.L. Chernous'ko, Yu.N. Andreev, N.A. Bobylev and V.M. Khametov for discussion and valuable suggestions. The author also wishes to thank E.A. Andreeva as well as to his coworkers N.L. Lepe, A.V. Babichev and V.I. Finyagina for their help in the preparation of the manuscript and for solving a number of problems. Of course, for any deficiency in the book the author is solely responsible.

Moscow, April 1984   A.G. Butkovskiy

Introduction

With the publication of Henri Poincare's celebrated treatise "On Curves Determined By Differential Equations" the idea of the phase portrait of a dynamical system became a powerful tool for local and global investigation of properties of dynamical systems, the base of the qualitative theory. Through the works of A.A. Andronov and his colleagues, pupils and followers, the phase portrait of the dynamical system was transformed into a working instrument for analysing and synthesising the set of mechanisms and systems from substantively diverse fields of physics, technology and other branches of science. This was made possible thanks to the simplicity and vividity (geometricity) of the phase portrait, which reveals not only the local pecularities of the system but represents the system globally providing a vivid picture of the behaviour of the dynamical system in the large.

For an uncontrollable dynamical system (UDS)

$$\dot{q} = f(q), \tag{1}$$

where q is an n-vector describing the state of UDS, the phase portrait is the aggregate (family) of phase trajectories in its phase space or, to use the customary phrase, in its state space. This portrait is indeed a portrait on which many facts are visible. This becomes possible thanks to the fact that through every point of the phase space (state space) there passes (by the uniqueness theorem) one and only one curve (trajectory) except for individual points and individual manifolds with dimension less than n. These are referred to as 'singular points" and "singular manifolds" respectively. This portrait is especially simple and expressive in the case of two-dimensional systems (n = 2) which can be represented on the plane or in a part of it or on two-dimensional manifolds (surfaces) having more complex structure such as a sphere and a torus etc.

In an UDS an important role is played by singular points and singular solutions. On these singular sets the existence and uniqueness theorems are violated. As a result, a whole family of trajectories may pass through a singular point.

From this point of view, for a controllable dynamical system (CDS) of the form

$$\dot{q} = f(q, u), \tag{2}$$

dealt with in the present book, every point of its phase space is, as a rule, a singular point. Here q denotes an n-vector describing the state of CDS and u denotes a control depending on time t. Thus if the representative point of (2) is located at a given point q of the phase space at the given instant, in the subsequent instant it can move along any trajectory belonging to a bundle of trajectories originating from q. The choice of a particular trajectory from the bundle is determined by the control u which is chosen at the given and subsequent moments of time.

Thus the first impression is that there is a chaos in multiple interweaving trajectories which fail to constitute a systematic figure. As a matter of fact, for an UDS at every point q of the phase space there is defined, as a rule, a unique direction of velocity (tangential to the trajectory). In contrast, for a CDS with every point q there is associated an entire cone containing the bundle of admissible directions of velocity (tangential to the trajectory). Thus, unlike an UDS which is determined by a field of directions (isoclines), a CDS is determined by a field of cones of admissible (feasible) velocity directions of CDS. Such a situation is equivalent to defining a CDS by a differential inclusion: the velocity of a CDS at the given point of the state space belongs to a cone of admissible velocity directions drawn at the point in question. What is more, the inclusion of limiting trajectories associated with the so called "sliding rules" in the set of admissible trajectories enables us to deal with a field of convex cones.

It is therefore necessary to first study and classify the convex cones. It turns out that every convex cone in the n-dimensional space with vertex at the origin is associated with its type a_n^m determined by two indices m and r. The index m is the dimension of the minimal linear subspace L^m containing the entire cone and r is the maximum number of linearly independent support planes at the vertex of the cone in L^m . This implies that in the general case of a CDS with an n-dimensional state space there can be in all $\dfrac{(n+1)\ (n+2)}{2}$ distinct types of cones. Of course, for a particular CDS not all types of cones may be present.

Now the state space of CDS can be decomposed into disjoint sets such that at all points of a particular set the type of the cone is saved. This enables us to deal with the field of only one type of cones on each of these sets. And this can significantly simplify the investigation regarding the nature of admissible trajectories lying in the given set or intersecting it.

The complexity involved in investigating the nature of admissible trajectories and the set of one type of cones depends on the form of this set, its size and on the type of the cone. In the general case here one comes across a wide range of various possibilities, some of them demanding very scrupulous and nontrivial discussion. In some cases the investigation is quite simple, at least in principle. We have in mind the case, for example, of a domain of free trajectories of a CDS or the case of a set of absolute equilibrium of CDS or that of a set having a field of tapered solid cones. Much more complicated to deal with are those cases where invariant manifolds of various dimensions are involved. This is because one has to investigate linear (differential) Pfaffian forms determined by linear subspaces L^m and support planes associated with the given type of cones. It is pertinent to note that although the given CDS may be prescribed initially in a linear n-dimensional space, yet due to the possible presence of an invariant manifold, which the representative point of CDS cannot leave along any admissible trajectory either in forward time or in backward time, one has to necessarily study a CDS on invariant manifolds.

The investigation of a CDS on sets where the type of the cones is saved can be called "local investigation". However, there also arises the question of "global investigation". This includes the possibility and nature of transition from one set of given fixed type to its neighbouring set. The picture of the global relation in CDS can be represented by a graph where the vertices represent the sets of cones of fixed type (or their subsets) and the edges indicate the possible transition of the representative point of CDS from one set to another.

The above picture representing the local and global properties of a CDS will be called its *phase portrait*. In a wider sense, by a phase portrait of a CDS we mean the aggregate of geometrical as well as analytical aids and notions which enable us to have a full and, as far as possible, vivid geometrical picture of local and global nature of the behaviour of admissible trajectories of a CDS. An individual notion or aid of the phase portrait will be known as an *element* of the phase portrait of the CDS.

Apart from the elements of phase portrait just listed, other notions introduced in the present book can also prove to be very useful. These include the notion of a "separating" hypersurface and that of an "admissible" surface in the state space of the CDS. A very interesting and useful element of the phase portrait, introduced in the present book,is the notion of the "boundary of trajectory (integral) funnel" of the differential connection which is equivalent to Eq. (2). The boundary of trajectory funnel exists for, at least, a two-dimensional CDS (on the plane and on two-dimensional manifolds) in a domain of the space which is distinct from the domain of unconstrained (free) trajectories. In a space whose dimension exceeds 2, the boundary of the trajectory funnel may not exist. The boundary of a trajectory funnel (if if exists) represents the lateral surface of a conoid with vertex (pointed) at a point q of the state space of the CDS. None of the admissible trajectories originating from q can continue beyond this conoid. To underline this fact we shall perform hatching on the external side of the conoid. It should be remarked that the boundary of the trajectory funnel constitutes a part, the lateral part, of the boundary of the domain of reachability from q. The "base" of this conical domain is the Bellman surface of the speed problem (for a given time $T > 0$).

In our view what is striking is that the boundary of the trajectory (integral) funnel of the differential inclusion of the CDS (in case it exists) is nothing but the characteristic conoid for the nonlinear partial differential equation of first order

$$ H\left(\frac{\partial z}{\partial q}, q \right) = 0 \qquad (3) $$

in the unknown scalar function z (q). It was found that the cone of admissible velocity directions of the CDS is the Monge cone of this differential equation. It is also discovered that the function H (p,q) in (3), which we shall call the *Hamiltonian* of the CDS, is a support function of the set of admissible velocities of the given CDS.

Since to Eq. (3) there corresponds a canonical system of ordinary differential equations of order 2n, the characteristic conoid of Eq. (3) and, hence, also the boundary of a trajectory funnel of CDS (2) can be constructed as a family of solutions of the Cauchy problem (and not the boundary-value problem!) for the canonical system with completely defined specific initial conditions.

It should be noted that the question of the existence of boundaries of trajectory (integral) funnels is closely related to that of the integrability of Pfaffian differential equations which describe the type of the cones of admissible velocity directions.

Of course, not all the questions connected with the construction of the phase portrait of a CDS in the large and its individual elements have been settled in the present book. In reality, the author has suggested only a programme for investigating a wide range of problems connected with the development and activisation of geometrical methods and aids for representing a CDS. And, of course, it is not suggested here that analytical and algebraic methods of investigation have to be neglected. These methods can and must supplement in equal measure the geometrical approach, and this reflects well the present tendency in the development of mathematical methods.

Especially simple, at least in principle, looks the phase portrait of the CDS on a plane (or on other two-dimensional manifolds). Indeed, in this case we have to deal only with a pair of extreme trajectories passing through a given point of the plane. Roughly speaking, the phase portrait of a two-dimensional CDS can be treated as a superposition of phase portraits of two UDSs. Thus, unlike the usual problem of the qualitative theory, where one has to study the phase portrait of a single UDS, in the investigation of a two-dimensional CDS we have to examine the interaction of two superposed families of usual phase trajectories obtained for two independent UDSs of type (1). It is clear that the study of this higher level problem must be based on the results and methods of investigating phase portraits of UDSs.

As in the case of the qualitative theory of an UDS, the study of a CDS begins by identifying the singular points and manifolds (on the plane they are, in general, singular curves). Apart from singular points, which are typical of an UDS, in the investigation of a CDS there arise its own singular manifolds. An important role is played here by the manifolds (curves) of reverse hatching. The singular sets of a CDS, like the typical singular sets of phase portraits of an UDS, play a decisive role in discovering the properties of the CDS. What is striking is that these singular sets (just as in the case of an UDS) do not require integration for their identification, and can be obtained by simpler, very often algebraic, operations.

The phase space and, in particular, the phase plane were used earlier also in the study of the local and global properties of a CDS. In this connection it is interesting to observe that the transition curves and surfaces in the phase space of system that are optimal in speed, which were first constructed by Feldbaum and Bushau, are, from the point of view of the phase portrait of the CDS introduced here, nothing but the boundaries of the trajectory funnels constructed in backward time and having vertices at singular points.

Although the identification of singular manifolds, as noted above, is relatively simple (it depends only on the form of the differential equation), the picture of possible motions of CDS in the neighbourhood of singular manifolds can be very wide ranging and complex. A "transition" through these singular manifolds can completely change the picture of hatched boundaries of trajectory funnels. The study of all types of these changes and coupling of phase portraits of various domains of the state space must constitute one of the major problems of the qualitative theory of controllable systems.

As in the case of an UDS, it is clear that with an increase in the dimension of the system the investigation of the phase portrait becomes very complicated. However, one can hope to do better in the study of concrete classes of CDS such as a linear, bilinear, analytic or coordinate-wise linear CDS or a CDS linear in control etc.

The phase portrait of a CDS, especially in the two-dimensional case, can prove to be very useful. Sometimes it can provide exhaustive solution to several basic control problems such as the problems of controllability, finite control, optimal control, stability or the synthesis problem subject to additional conditions (say, on the phase coordinates).

For instance, in solving the complicated problem of optimal control where it is required to minimise a function of the final point of the phase trajectory ϕ (q(T)) subjected to additional phase conditions a very useful device is to superpose on the phase portrait the level surfaces of the function and the contours of the phase constraints. This superposition enables us to determine visually the admissible optimal trajectory.

As noted earlier, the problem becomes very complex when we move to the three dimensional case. Here we need to study the interaction among an infinite set (linear continuum) of families of trajectories and velocity fields. For a system of order n, we must study, in general, ∞^{n-2} families of such fields and their interaction. In recent years, the study of interaction of the set fields is being intensively carried out by algebraic methods in the adequate terms of sets of one-parameter groups and Lie algebras (or multi-parameter groups and algebras) corresponding to the given CDS.

Of course, in its full extent this problem is as difficult as, and in the light of above remarks much more difficult than, the corresponding problem regarding the phase portrait of an UDS. But in our view one can already proceed (this has been partially done in the present book and some other works) to develop the general theory of the phase portrait of a two-dimensional CDS on the plane and on two-dimensional manifolds. We can also hope to succeed in solving this problem for arbitrary (at least, finite-dimensional) linear and bilinear CDSs as well as for nonlinear CDSs, say, with a linear control. We can expect to have a relatively quick success in solving this problem for CDSs described by smooth or analytical functions. Such an approach can give a clear geometric interpretation in a concentrated form to several results obtained earlier by means of group-theoretic and algebraic methods.

The notion of the phase portrait of a CDS seems to us to have a bright prospect also because the phase portrait of a CDS can be constructed, in principle at least, by means of a computer. This is already being done and a few results have been obtained in this direction. Concerning the construction of the boundaries of integral funnels occurring in the phase portrait of a CDS with n > 2, we do not have any other method for their vivid investigation apart from the method of sections and the method of projections onto subspaces of lower dimensions. These operations of constructing sections and projections can also be entrusted to a computer, and here new vistas for activities are opened.

It is also interesting that the partial differential equation (3) can be interpreted as the Hamilton-Jacobi equation for a mechanical (uncontrollable) system which is generated in a natural manner by the original CDS. In with n > 2, we do not have any other method for their vivid investigation apart from the method of sections and the method of projections onto subspaces of lower dimensions. These operations of constructing section and projections can also be entrusted to a computer, and here new vistas for activities are opened.

It is also interesting that the partial differential equation (3) can be interpreted as the Hamilton-Jacobi equation for a mechanical (uncontrollable) system which is generated in a natural manner by the original CDS. In these terms, the hatched conoid or the hatched boundary of the trajectory funnel for the associated differential connection can be interpreted as the projection of the Lagrange manifold into the configuration space of the mechanical system thus obtained. It turns out that the boundary of the trajectory funnel, which in reality is the boundary (or its part) of the domain of accessibility from the given point, is woven from the characteristics and characteristic strips of the corresponding Hamilton-Jacobi equation and the Hamilton canonical system.

What is striking in our view is that there also exists the reverse connection. For example, to every uncontrollable mechanical system of dimension 2n with a non-homogeneous Hamiltonian there corresponds a completely defined controllable system of dimension n+1 which had generated the original/mechanical system. Such relationships after they have been explicitly followed through may not seem to be surprising. But the crux of the matter is that in any event they make it possible to apply the powerful and well developed tool available for investigating uncontrollable systems to the investigation of controllable systems. Of course, also useful is the influence exerted the other way round. This is perhaps one of the paths that can lead us to a general theory of dynamical systems, uncontrollable as well as controllable.

A few words should be said regarding the connection between the phase portrait of a CDS and the notion of a continuous medium and processes flowing into it. It turns out that to every CDS of type (2) there corresponds a continuous medium, for example optical medium, where a certain perturbation is propagated. Conversely, to every such optical medium there corresponds a CDS of type (2). This correspondence is a natural consequence of the one-to-one correspondence between controllable and uncontrollable mechanical systems mentioned above.

Another, perhaps less obvious, analogy with the processes in continuous media and systems with distributive parameters follows from the fact that the characteristics of the original controllable systems of type (2) can be interpreted as material relationships for the processes of propagation of substance (for example, of heat, matter and energy). This enables us to associate with the controllable system of type (2) an operator, the Laplace operator for instance, describing the processes of propagation of substance in the medium. What is interesting is that with the usual classical media, where, for example, heat is propagated, there are associated very simple controllable systems of type (2) and vice-versa. To more complex systems of type (2) there correspond more complex continuous media with intricate nonlinear properties and the presence of internal activities. Such continuous media are usually created artificially. They are playing an increasing role in diverse fields such as the construction of composite materials, synthesis of active distributed regulators for stabilization and control of complicated objects with distributed parameters [22-30,64,72,99).

We also note the connection between the theory of a CDS of type (2) and that of a CDS with distributed parameters [22-30,72] governed by partial differential equations. This connection consists in that Eqs. (2) yield the equations of characteristics (bicharacteristics) of the corresponding partial differential equations. One of the first problems of the theory of control of distributed parameter[*] was solved by using this simple idea [22,23,27]. However, it is regrettable that after more than twenty years this approach did not receive any theoretical development. The author hopes that the method of phase portrait proposed in the present book will stimulate new investigations in this direction and will lead to new useful results, both theoretical and practical in the field of control, particularly of distributed plants.

---

[*]) We mean the optimal control problem of heating materials in heating mechanisms. This problem is vital from the practical point of view.

2

Controllable Dynamical System (CDS)

By a CDS we shall mean, in this book, a system governed by the equation

$$\dot{q} = f(q, t, u), \tag{1}$$

where q is a column vector (point), with components $q^1, ..., q^n$, in an n-dimensional space { q }. We call this space the *state space* of the CDS or the *phase space* of the CDS. By { q } we shall mean, as a rule, the linear space R^n. If the phase space of the CDS (1) is an n-dimensional manifold, it will be denoted by M^n. Of course, the CDS (1) may be prescribed a priori on a manifold; for example, on a sphere or on a torus etc. However, as we shall see later (Sec.19), a manifold as a phase space of the CDS (1) can also arise a posteriori; for instance, when CDS (1) has an invariant manifold M^m immersed in R^n ($m \leq n$). Further, u will denote a *value of the control* which it assumes from an arbitrary given non-empty set U, *known as the set of admissible values of the control*. Sometimes u is also referred to as a *parameter of control* or a *controlling parameter*. The *velocity* $\dot{q}$ of a CDS is a column vector (point), with components, $\dot{q}^1, ..., \dot{q}^n$ in an n-dimensional affine space { q }. This space is known as the *velocity space* of the CDS or the *tangent space*, and is also denoted by T_q { q }. The origin in the tangent space q at the point q will be denoted by $\dot{o}$ (q). The independent scalar variable t will denote time and will assume values from a segment [t_0, t_1] (possibly, unbounded) of the number axis t. The vector function f (q, t, u), with components $f^1, ..., f^n$, is assumed to be defined for any value of its arguments from the above mentioned sets.

It should be remarked that all the results obtained in the present book remain valid, unless otherwise stated, when U = U (q, t), (q, t) ∈ { q, t }, that is, when U depends on the state q and time t.

Apart from the above quantities, we shall also consider the *impulse* p of a CDS. It is a row vector, with components $p_1, ..., p_n$, belonging to the n-dimensional impulse space { p } = R^n (the cotangent space) of the CDS phase space. We shall also make use of the (n+1) – dimensional/vector space { q, t } consisting of column vectors (points) with components $q^1, ..., q^n$,t and call this space the *event space* of the CDS. The state q will also be referred to as the *representative point* of the CDS in the state space {q}.

9

In order to determine the unique solution of system (1), that is, to determine the function q (t), $t_0 \leq t$, it is necessary to prescribe the initial instant of time t_0 and the initial state q_0 of the CDS at $t = t_o$, that is

$$q(t_0) = q_0.$$

It is also necessary to determine the control of the CDS. This is given by a function u (t) defined on the segment $[t_o, t_1]$ and assuming values in U. It is assumed that the control u (t), $t_0 \leq t \leq t_1$ belongs to a definite class of functions A $[t_o, t_1]$ defined on the time segment $[t_o, t_1]$; for example, to the class of measurable or piecewise continuous functions defined on $[t_o, t_1]$. Such a control u (t) $\in$ A $[t_o, t_1]$, u $\in$ U will be called an *admissible control*.

We assumed that for a given admissible control u (t) and a given initial condition (2), Eq. (1) has a unique and, at least, absolutely continuous solution q (t) defined on the same time segment $t_0 \leq t \leq t_1$ and satisfying the initial condition (2). That is, we assumed that a unique q (t) exists almost everywhere on $[t_o, t_1]$ [88]. This assumption is fulfilled if, for example, (i) the set of admissible values U is a set of R^n, (ii) the components of the vector f (q, t, u) in Eq. (1), that is, the functions f^i (q,t,u), i = 1, ...,n, are jointly continuous in the variables a q^1, ..., q^n, t, u, and are continuously differentiable with respect to q^1, ..., q^n, A $[t_o, t_1]$ is a class of piecewise continuous vector functions.

Further, the function q (t), $t_0 \leq t \leq t_1$ which is the solution of Eq. (1) for some admissible control u (t), $t_0 \leq t \leq t_1$, will be called an admissible motion of the CDS corresponding to the control u(t). Thus, by definition, to every admissible motion q(t), $t_0 \leq t \leq t_1$ there corresponds at least one admissible control u (t), $t_0 \leq t \leq t_1$, the action of which, in fact, results in this motion. The graph of the admissible motion q (t), $t_0 \leq t \leq t_1$, in the event space { q, t } will be called the *admissible integral curve*. The projection of the admissible integral curve q (t), $t_0 \leq t \leq t_1$, onto the state space {q} will be referred to as the *admissible trajectory* directed from the initial point q (t_0) = q_0 to the final point q (t_1) = q_1 and will be denoted by q(q_o, q_1). The direction along the trajectory from its initial point to the final point will be taken as the positive direction.

The motion of a CDS in its state space { q } is governed by the motion of its representative point along the positive direction of the trajectory. And, in its turn, the motion of the representative point q is uniquely determined by the magnitude and the direction of the velocity vector q, which is directed along the tangent at the point q. Here one must state that in general some of the solutions q (t) of relation (2) on $[t_o, t_1]$ may correspond to controls u(t) that do not lie in A (t_o, t_1), the original class of admissible controls of the CDS (1). As regards the uniqueness of the solution, we assume that for corresponding f (q, u), u $\in$ U and Φ (q)[*] two distinct solutions q_1 (t) and q_2 (t) satisfying (2) with a common initial point q_o generate significantly distinct controls u_1 (t) and u_2 (t) for CDS (1).

---

[*] For the definition of F(q), see Sec.3 - Tr.

3

Differential Inclusion. Equivalence Classes of CDS

We assume that the given CDS is autonomous, that is, it is governed by an equation of the form

$$\dot{q} = f(q, u), \quad u \in U(q), \quad q \in \{q\}, t > o, \quad q(o) = q_0. \tag{1}$$

We fix the point $q \in \{q\}$ and let u run through the entire set $U(q)$, which possibly depends on q as well. Then, in view of (1), the point $q \in \{q\}$ runs through a set $\Phi(q)$ in the space $\{q\}$. We denote this set by $f(q, U)$ and call it the *set of admissible velocities* of the given CDS. We assume that $\Phi(q)$ is a non-empty closed set. Thus we obtain the set $f(q, U) \equiv \Phi(q)$ by mapping U into the velocity space $\{\dot{q}\}$ by means of the function $f(q, u)$ for a fixed $q \in \{q\}$.

The notion of the set of admissible velocities of a CDS enables us to replace Eq. (1) governing the CDS by an equivalent expression in terms of a *differential inclusion*:

$$\dot{q} \in f(q, U) \quad \text{or} \quad \dot{q} \in \Phi(q). \tag{2}$$

By a solution of the inclusion (2) with the initial condition $q(o) = q_0$ we mean an absolutely continuous vector function $q(t)$, $\theta \le t \le T$, such that the following condition is satisfied almost everywhere on the segment $[O,T]$: the point $\dot{q}(t)$ belongs to the set $f(q(t),U)$, that is,

$$\dot{q}(t) \in f(q(t), U), \quad \text{or} \quad \dot{q}(t) \in \Phi(q(t)), a \forall t \in [O, T].^* \tag{3}$$

We call the solution $q(t)$, $O \le t \le T$, the *admissible motion* of the CDS *governed by inclusion* (2). If we know a solution $q(t)$ of the differential inclusion (2), we can recover that control $u(t)$ for system (1) which generates this solution [14, 111, 112, 120].

---

*) The notation a $\forall$ stands for "for almost all".

11

The equivalence of two modes of describing a CDS, one in terms of an equation and the other in terms of a connection, means that the set of all admissible motions of CDS (1) and that of CDS described by inclusion (2) coincide, unless stated otherwise. On the other hand, however, it is also evident, and examples can be easily cited in support of this fact, that distinct CDS (1) may be equivalent to one and the same connection (3). In this way all the equations of CDSs, and by the same token CDSs themselves, are divided into equivalence classes. To one class belong all those CDSs which have the same connection (2), that is, which have the same set f(q, U). In other words, to each set f(q, U) there corresponds a certain "enlarged" set of CDSs of form (1).

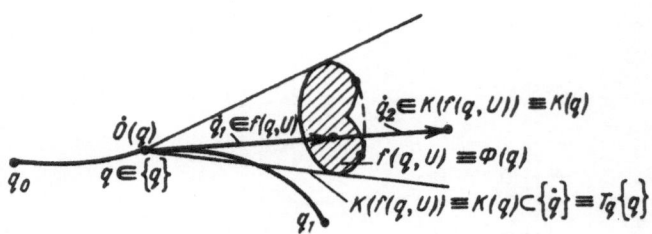

Fig.3.1.

All systems of form (1) belonging to the same class will be referred to as *equivalent* CDSs.

Further steps in the enlargement of the set can be taken in the following two directions. (i) We more over from inclusion (2) to the connection

$$\dot{q} \in K (f (q, U),\qquad\qquad\qquad (4)$$

where $K (f (q,U))$ denotes the cone consisting of all those rays originating from the point $\dot{O} (q) \in \{ \dot{q} \}$ which have a non-empty intersection with $f(q,U)$, other than the point $\dot{O}$ (q) if $\dot{O}$ (q) $\in$ f (q,U). If f(q,U) consists of the single point $\dot{O}$ (q), we assume that K (f(q,U)) coincides with this point. The cone K (f(q,U) will be called the *cone of admissible directions* of the CDS. (ii) From connection (2) we switch over to the inclusion

$$\dot{q} \in \text{conv } f(q,U),\qquad\qquad\qquad (5)$$

where conv $f(q,U)$, denotes the convex hull of the set $f(q,U)$.

Let us first discuss the step outlined in direction (i). We note that no matter what admissible trajectory of (4) is taken, to it there corresponds an exactly similar trajectory of (2) and vice-versa. This situation is clearly illustrated in Fig.3.1. It is clear from the figure that whatever be the velocity $\dot{q}_2 \in K$ (q), $| \dot{q}_2 | \neq O$, there exists always a velocity $\dot{q}_1 \in \Phi(q)$, $|\dot{q}_1| \neq O$, such that $\dot{q}_2$ and $\dot{q}_1$ are collinear and are in the same direction. Thus the replacement of connection (2) by inclusion (4) does not change the set of admissible trajectories of inclusion (2). We may say that in this change the geometrical information is preserved. However, in this transition from (2) to (4) the information regarding possible magnitudes of velocities in motion of the representative point along an admissible trajectory is lost, that is, the time (kinematic) information is lost.

Analytically, the transformation from (2) to (4) can be effected by introducing an additional control in the form of a scalar function α (q) ≥ 0. Then (4) is equivalent to the inclusion.

$$\dot{q} \in \alpha \, (\, q \,) \, f(q, \, U). \tag{6}$$

We can then get rid of α (q) by changing the time scale, that is, by introducing a new parameter τ in place of τ by the formula

$$d\tau = \alpha \, (q) \, dt.$$

Thus the cone K(f(q, U)) defines a class of equivalent (in the aforementioned sense) CDSs (1) and (2). To this class belong all CDSs (1) and (2) having one and the same cone K(f(q, U) of admissible direction of velocities. CDSs (1) and (2) equivalent in this sense will be referred to as *trajectory-wise equivalent* systems.

We now examine the steps in the enlargement in direction (ii). This step is much more vital than step (i), and has been investigated in several works [14, 37, 11, 112, 120]. The matter is that connection (5) can contain some admissible trajectories which are not "ordinary" admissible trajectories of (2) or of the original Eq. (1). However, it can be established (see, for example, [120]) that there exists a sequence of admissible controls u_i (t) of Eq. (1) in A(O,T), the class of piecewise continuous functions, not converging to any ordinary function but for which the corresponding sequence of motions q_i (t) converges to an absolutely continuous function q(t) satisfying inclusion (5). Such solutions q(t) of (5) are called *generalized admissible motions* of the CDS *in the space* {q,t} and the corresponding trajectories are known as the *generalized admissible trajectories* of the CDS in {q}.

Thus, having regard to generalized admissible motions, we can treat inclusion (2) and (5) as equivalent. It is clear that there can be distinct CDSs of type (2) which belong to the same equivalence class determined by (5). CDSs (2) and (5), *equivalent in this sense*, will be referred to as equivalent in the generalized sense.

Finally, we move to a still larger equivalence class. From inclusions (4) and (5), we move to the inclusion[*]

$$\dot{q} \in \text{conv} \, K \, (\, f \, (\, q, \, U \,) \,) = K \, (\, \text{conv} \, f \, (\, q, \, U \,) \,). \tag{7}$$

It is easily proved that [97]

$$\text{conv} \, K \, (\, f \, (\, q, \, U \,) \,) = K \, (\, \text{conv} \, f \, (\, q, \, U \,),$$

that is, the operation of constructing a convex hull and that of constructing a cone are commutative. Thus we have found the largest (among those considered here) equivalence

---

[*] In the most general case the transformation to this set must be effected more accurately in order to guarantee the existence of solutions of inclusion (4), (5) and (7). However, in many cases it suffices to construct a convex set straightaway, as done here.

class, which is characterized by the set K(conv f(q, U) or, what is same, by the set conv K(f(q, U). We call this class the *equivalence class over generalized trajectories.*

Without loss of generality, we shall assume in the sequel, unless otherwise stated, that f(q, U) is a non-empty convex set and that the corresponding cone K(f(q, U) is also convex; this cone will be denoted by K(q).

Apart from the cone K(q), an important role in our discussion will be played by its conjugate (dual) cone $\overline{K}(q)$ (see Fig. 14.1):

$$\overline{K}(q) = \{p \in R^n \mid p\dot{q} \le 0, \forall \, \dot{q} \in K(q)\}. \tag{8}$$

The successive enlargement of equivalence classes discussed above is presented schematically in Fig.3.2.

We note that notion of the equivalence class introduced here satisfies the usual requirement of transitivity, reflexivity and symmetry [3]. We also remark that many problems which do not formally constitute problems on differential inclusion can be reduced to them [14] (to problems in differential inequalities, for instance).

Since a non-autonomous system can be reduced to an autonomous system (Sec.3) by introducing an additional (n+1)th coordinate $q^{n+1} \equiv t + t_0$, all that has been stated in the

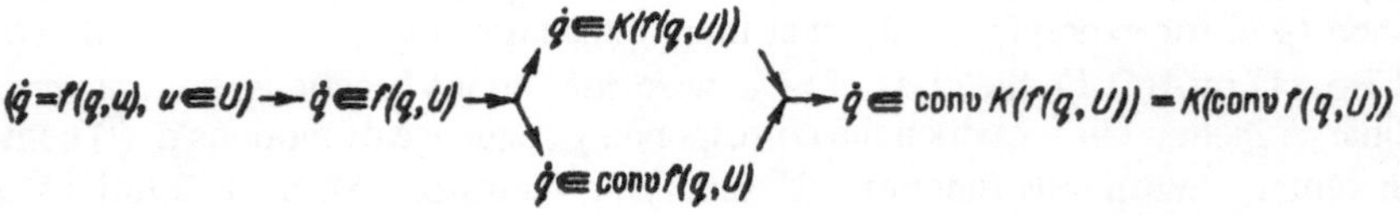

$$(\dot{q}=f(q,u), \; u \in U) \rightarrow \dot{q} \in f(q,U) \rightarrow \left\{ \begin{array}{c} \dot{q} \in K(f(q,U)) \\ \\ \dot{q} \in conv\,f(q,U) \end{array} \right\} \rightarrow \dot{q} \in conv\,K(f(q,U)) = K(conv\,f(q,U))$$

Fig.3.2.

present section remains valid for a non-autonomous system too. Indeed, suppose that the CDS

$$\dot{q} = f(q, t, u), \qquad u \in U(q, t) \tag{9}$$

is non-autonomous. Then (Sec. 3) Eq.(9) can be expressed in the form

$$\dot{\overline{q}} = \overline{f}(\,\overline{q}, u), \, u \in U(\,\overline{q}) \tag{10}$$

and the corresponding connection will be of the form

$$\dot{\overline{q}} \in \Phi(\,\overline{q}) \tag{11}$$

The state space for the autonomous CDS (10), (11) coincides with the event space for the non-autonomous CDS (9), that is,

$$\{q, t\} = \{q, q^{n+1}\} = \{\,\overline{q}\,\}, \, q^{n+1} = t + t_0 \, \dot{q}^{n+1} = \dot{t} = 1.$$

Therefore in the tangent space $T_{\bar{q}}\{\bar{q}\}$ the set $\Phi\{\bar{q}\}$ will constitute a *plane* figure and will wholly lie in the plane $\dot{q}^{n+1} = \dot{t} = 1$. This implies that the cone $K(\Phi(\bar{q})) = K(q, t)$ containing $\Phi(\bar{q})$ *can never coincide* with the entire space.

It should be remarked that if the system (9) is autonomous all the same, the cone $K(q, t)$ is independent of t in the space $\dot{q} = \left\{ \overset{\pm}{\dot{q}}, \dot{t} \right\}$. In this case, the cone $K(q)$ lying in the space $\{\dot{q}\}$ is the projection of the cone $K(q, t)$ onto the space $\{\dot{q}\}$

To conclude the present section, we remark that in spite of the fact that $\Phi(q)$ may be a closed set the corresponding cone $K(q)$ may not be closed. This fact is illustrated in Fig.3.3. Such cases require special investigation. We assume in what follows that the cone $K(q)$ contains all its limit points except perhaps the vertex $\dot{O}(q)$.

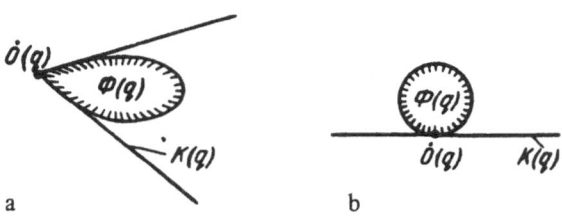

Fig. 3.3.

Transformation of CDS to Phase Velocity Unit Vector

In many problems that arise in the study of a CDS what interests us in the space {q} is the trajectory itself and not the law of motion of the representative point q along this trajectory. This is the case when, for example, one has to deal with controllability problems. In this case the existence of the trajectory joining, say, two given points has to be guaranteed. But the law of motion itself of the representative point along the trajectory during this time is not always of interest to us. Of course, here it has to be ensured that the time taken to move from the initial point to the final point is finite. In this case it proves more convenient to transform the original CDS to a new CDS with unit velocity vector: $|\dot{q}| = 1$. In effect, this transformation performs a "separation of variables" characterising the form of the trajectory in {q} on one hand and, on the other, the law of motion along the trajectory during this time.

The original CDS

$$\dot{q} = f(q, u) \tag{1}$$

can be transformed in the following manner. We divide both sides of (1) by $|f(q, u)|$, assuming that for any point $q \in \{q\}$ and $u \in U$ there is a quantity $\varepsilon > 0$ such that. Then we have

$$\frac{dq}{|f(q, u)| dt} = \frac{f(q, u)}{|f(q, u)|} \tag{2}$$

We introduce a new independent variable τ by the formula

$$d\tau = |f(q,u)| dt,$$

$$\tau = \int_0^t |f(q(\theta), u(\theta))| d\theta \tag{3}$$

16

in place of t and treat the (dependent) variables q and u as functions of τ :

$$q = k(\tau), u = \upsilon(\tau) \tag{4}$$

with

$$q = q(t) = k\left(\int_0^t |(q(\theta), u(\theta))| d\theta\right) \tag{5}$$

$$u = u(t) = \upsilon\left(\int_0^t |f(q(\theta), u(\theta))| d\theta\right) \tag{6}$$

Then (2) assumes the form

$$\dot{k}(t) = \frac{f(k\tau\upsilon(\tau))}{|f(k(\tau), \upsilon(\tau))|} \equiv \varphi(k(\tau), \upsilon(\tau)), \upsilon \in U \tag{7}$$

Where $|\varphi(k,)| = 1$.

The inverse transformation of (3) is given by the formulae

$$dt = \frac{d\tau}{|f(u(\tau), \upsilon(\tau))|} \quad \text{or} \quad t = \int_0^\tau \frac{d\theta}{|f(k(\theta), \upsilon(\theta))|} \tag{8}$$

The significance of transformation (3) lies in that in the new system (7) the trajectories in the space {q} coincide with those of the original systems (1) in the same {q} and the direction of motion along the trajectory is preserved. The velocity vector is of unit magnitude: $|\dot{k}(\tau)| \equiv 1$. This transformation from (1) to (7) clearly amounts to introducing a (variable) time scale or, as it is customary to say, to introducing a new parameter τ in place of t by means of the formula (3), the direction of motion along the trajectory being preserved (that is, the orientation of the trajectory remains unchanged).

In discussing the time dependent characteristics of the original Eq. (1) with the aid of the transformed system (7) we must make use of formulas (5) and (6), that is, we must go back from τ to t using formulas (3) in which only one nonnegative scalar function $|f(q, u)|$ appears and only one integration is performed.

Thus we find that any CDS is characterised by Eq. (7), with $|\dot{k}|$ describing the form of the trajectory in {q}, and a nonnegative scalar function describing the law of motion of the representative point along the trajectory determined by Eq. (7). We can assume that $|\dot{q}| = |f(q, u)| = 1$ in Eq. (1) at the point where $|f(q, u)| \neq 0$.

The above arguments enable us to standardize the set of admissible velocities of an arbitrary CDS. In the space {$\dot{q}$}, this set is determined by a cone bounded by a unit sphere. Thus all the admissible velocities are located in a certain *solid angle*, and the end points of these vectors lie on the unit sphere. Here we have, in fact, performed a central projection of the convex set $f(q, u)$ onto the unit sphere in the space {$\dot{q}$}.

Indicatrix of CDS

We have already stated (Sec. 3) that throughout the book by the set $\Phi\ (q)$ of admissible velocities of a CDS or, what is same, by $f(q, U)$ we mean a *convex set*. The members of this set are the points of the tangent space $T_q\ \{\ q\ \} = \{\ \dot{q}\ \}$. In the investigation of a concrete CDS an important part is played by the properties of $\Phi\ (q)$ and the associated cone $K\ (q)$. In particular, of great significance are the dimensions of $\Phi\ (q)$ and $K\ (q)$ as well as the location of $\Phi\ (q)$ with respect to the origin $\dot{O}\ \{q\} \in \{\dot{q}\}$. A few examples illustrating the location of these sets are given in Fig. 5.1.

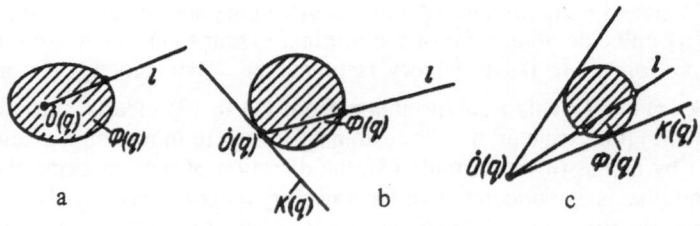

a   b   c

Fig. 5.1

Suppose that the dimension of $\Phi\ (q)$ is m : dim $\Phi\ (q) = m$. If $m = n$, the boundary $\partial\ \Phi\ (q)$ is a hypersurface in the n-dimensional space $\{\dot{q}\}$. This hypersurface is known as an *indicatrix* of the CDS. Suppose that the equation of an indicatrix is

$$\sigma\ (\dot{q}, q) = 0, \tag{1}$$

Where $q \in \{q\}$ plays the role of a parameter because this surface can naturally change with q at which the surface has been constructed. We assume that the domain of admissible velocities $\Phi\ (q)$ is defined by the inequality $\sigma\ (\dot{q}, q) \le 0$.

18

If $m < n$, the smallest dimension of the linear (affine) manifold containing $\Phi(q)$ wholly is also m. Such a linear manifold of smallest dimension m will be referred to as the *underlying minimal linear manifold of controls* of the CDS and denoted by L_m*).

In particular, if $\dot{O}\{q\} \in L_m$ then L_m is a linear subspace of $\{\dot{q}\}$ and in this case it is denoted by L^m.

The manifold L_m can be defined in a number of ways. For example, we can take a system of $m + 1$ *affinely independent* vectors $b_0, b_1, \ldots, b_m$. Here $b_1 - b_0, \ldots, b_m - b_0$ are linearly independent. In this case all the points lying in L_m, and only they, can be expressed in the form

$$\lambda_1 (b_1 - b_0) + \ldots + \lambda_m (b_m - b_0) b_0, \tag{2}$$

or

$$\lambda_0 b_0 + \lambda_1 b_1 + \ldots + \lambda_m b_m, \ \lambda_0 + \lambda_1 + \ldots + \lambda_m = 1. \tag{3}$$

The quantities $\lambda_0, \lambda_1 + \ldots + \lambda_m$ are known as the *barycentric coordinates* in L_m. The boundary $\partial \Phi(q)$ lying in L_m together with $\Phi(q)$ can be described by the equation

$$\sigma(\lambda, q) = 0, \tag{4}$$

where $\lambda = (\lambda_1, \ldots, \lambda_m)$ and the inequality $\sigma(\lambda, q) \leq 0$ describes the set $\Phi(q)$ in L_m.

Such a description enables us to represent the vector $\dot{q}$ as a direct sum of the vectors $\dot{q}'$ and $\dot{q}''$ of size m and $n - m$ respectively, that is, $\dot{q} = (\dot{q}', \dot{q}'')$. The equation of the original CDS is given by the system

$$\dot{q}'' = f_1(q), \tag{5}$$

$$\dot{q}' = f_0(q, \lambda), \tag{6}$$

where the vector λ denotes the control. The striking feature here is that Eq. (5) (that is, the set of $(n - m)$ equations) does not depend on the control; Eq. (6) depends on the vector λ which plays the role of the control, the vector also having size m. The new control λ is subjected to the sole condition $\sigma(\lambda, q) \leq 0$.

If by another change of variables, where λ is replaced by μ, it is possible to represent the equation of the CDS in the form

$$\dot{q}'' = f_1(q) = (q', q''), \tag{7}$$

$$\dot{q}' = \mu, \tag{8}$$

---

*) In the general case, L_m is known as the *affine hull of the set* $\Phi(q)$ [95] or as the *carrier plane*.

where the control μ is of size m and subjected to the sole condition $\sigma_1 (\mu, \dot{q}) \le 0$, then the representation of the CDS in the form (5), (6) or (7), (8) can prove to be very convenient in investigating a complex CDS because the representation decomposes the original CDS in a definite sense.

What is more, if the function is $f_1(q', q'')$ is independent of q', this means that the CDS is decomposed into two independent systems

$$\dot{q}'' = f_1 (q''), \tag{9}$$

$$\dot{q}' = \mu, \tag{10}$$

one of which, namely (9), is an uncontrollable dynamical subsystem in the sense that no controlling actions take place in this system. At the same time, CDS (10) is independent of (9) with its own state space $\{q'\}$ that is independent of the state space $\{q\} = \{q', q''\}$ of the original CDS in the sense that clearly no admissible control μ can force q' to leave $\{q'\}$. But, unfortunately, such a decomposition takes place, in general, only locally because $L_m (q)$ and its dimension m depend on q. In the particular case where L_m and its dimension are constant and are independent of q (such cases are not rare and take place, for example, for linear systems) there arises the problem of investigating the controllability of CDS (10) in its state space $\{q'\}$ of dimension m.

However, in the general case where L^m and m do depend on q, there arises the question of splitting the original space $\{q\}$ into subsets on each of which m is constant.

Thus, when dim Φ (q) = m < n, the indicatrix of the CDS is not a hypersurface since its dimension is less than n – 1. However, when L^m is a subspace having minimal dimension and containing Φ (q) wholly, we can speak of the indicatrix as a hypersurface but now with respect to L^m. Therefore the equation of this "relative" hypersurface can again be written in form (1), that is, in the form $\sigma (\dot{q}, q) = 0$. However, in this case the points $\dot{q}$ are no longer free and must lie in L^m. The subspace L^m can be expressed as a system of linear algebraic equations in $\dot{q}$, that is,

$$\pi \dot{q} = 0, \tag{11}$$

where π is an mxn matrix defining L_m. The entires of π depend, in general, on q (see also Sec.8).

Thus, when dim Φ (q) = n, the indicatrix is defined by a single Eq. (1). When dim Φ (q) = m < n, the indicatrix is defined, in general, by a system of equations consisting of the nonlinear Eq. (1) and the linear Eq. (11) with an m x n matrix.

In Sec.7, we shall examine an alternative and very important method of describing an indicatrix which can be termed "dual" of the present method: it is based on the construction of a support function of the set f(q, U).

6

Degree of Freedom in Control of CDS

It makes sense to discuss the notion of the "power" of controlling effects of a given CDS depending on what "degrees of freedom" are acquired by the CDS under the action of admissible controls. This question has two aspects - local and global. In the local case we are concerned with the neighbourhood of that point q which represents the state of the CDS at a given time t and at moments close to t. The global aspect of the degree of freedom (or power) in control of the given CDS is connected with the global picture of controllability over the entire phase space {q} (or over its given part) of the CDS, and can be characterized in terms of the global phase portrait of the CDS. This question will be discussed later (Sec. 25).

We try to obtain a local characterization for the power of control of the CDS depending on the point $q \in \{q\}$. Intuitively, the degree of freedom is connected with the notion of "flexibility of control", that is, with the possibility of a change in the state of the CDS with a change in its velocity and a change in its direction, including, say, a reversal in the direction of motion of the representative point.

When the CDS is represented by Eqs. (3.1) and connection (3.2) this flexibility must be obviously characterized in terms of the dimension of the set f(q, U) and the dimension of the cone K(q). This flexibility must also depend on the fact whether or not the point $\dot{O}$ (q) belongs to the set f(q, U).

It is natural to assume that for a given f(q, U) the flexibility (degree of freedom in control) is maximum when the point $\dot{O}$ (q) $\in$ ri f(q, U), where ri M denotes the relative interior of the set M [97]. That is, when the cone K(q) coincides with an m-dimensional linear subspace L^m of the velocity space $\{\dot{q}\}$ (m ≤ n and n is the size of the vector $\dot{q}$). Then the integer m can be regarded, by definition, as the degree of freedom of control for the given CDS. We shall write degf u = m. This quantity provides a characterization for the power of control or the flexibility of control.

Fig. 6.1 depicts for the case dim $\{\dot{q}\}$ = 3 an ordered evolution in the degree of freedom of control as it diminishes from the possible maximum. The last two diagrams in Fig. 6.1

21

depict the case of a complete degeneration in control and the reduction of the CDS into a totally uncontrollable system (at the given point q). The case (j) in Fig. 6.1 portrays the state of rest (the statical equilibrium of the system).

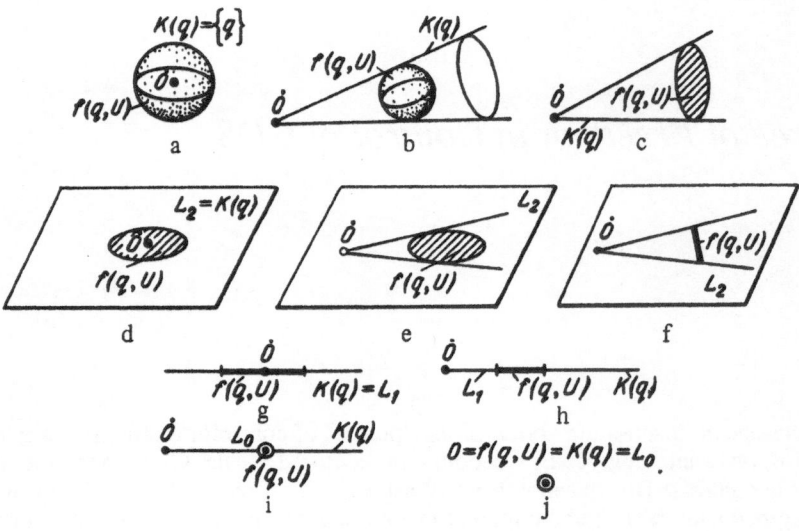

Fig. 6.1

To every sequence of evolution, such as above, we can ascribe a (not necessarily unique) suitable quantity for measuring the degree of freedom of the control. This characterization can be in terms of a numerical quantity or even a vector quantity. If a numerical characterization is preferred, then for the given n the set of numbers $\{\rho_i\}$ describing the degree of freedom of control must constitute an increasing (decreasing) finite sequence $\rho_0, \rho_1,..., \rho_r$ This sequence provides a characterization for all qualitatively distinct possible cases of feasible degrees of freedom in the given CDS of order n. For instance, we can introduce numerical characterizations ρ's for cases depicted in Fig. 6.1 as follows:

$$
\begin{array}{ll}
\text{(j)} & \rho_0 = 0; \\
\text{(i,h,g)} & \rho_1 = \frac{1}{2},\ \rho_2 = 1,\ \rho_3 = 1 + \frac{1}{2}; \\
\text{(f,e,d)} & \rho_4 = 2 - \frac{1}{2},\ \rho_5 = 2,\ \rho_6 = 2 + \frac{1}{2}; \\
\text{(c,b,a)} & \rho_7 = 3 - \frac{1}{2},\ \rho_8 = 3,\ \rho_9 = 3 + \frac{1}{2};
\end{array}
\qquad (24\ \text{A})
$$

and so on. The last case with $\rho = n + \frac{1}{2}$ corresponds to the maximum flexibility or "superfreedom" (see Sec. 9 concerning the domain of unconstrained (free) motions).

In the present notation ρ's are not numbers in the ordinary sense. They should/rather be regarded as numerical symbols since, for example, $\rho_3 = 1 + \frac{1}{2}$ and $\rho_4 = 2 - \frac{1}{2}$ and must be distinguished from each other. We can introduce the vector ρ. The, for example, the vectors $\rho_3 = (1 + \frac{1}{2})$ and $\rho_4 = (2 - \frac{1}{2})$ and are distinct in the usual sense, and thus reflect the presence of distinct degree of freedom in control u of two given CDSs.

In a somewhat crude classification of feasible degrees of freedom of control, we may not distinguish between the cases (h) and (i), (f) and (e), and (b) and (c). This is due to the fact that the noted pairs of CDSs will have the same set of trajectories in, at least, the neighbourhood of q. Then the table of values of ρ assumes the following form:

(j)	$\rho_0 = 0$;
(i,h)	$\rho_1 = \frac{1}{2}$;
(g)	$\rho_2 = 1$;
(f,e)	$\rho_3 = 1 + \frac{1}{2} = \frac{3}{2}$;
(d)	$\rho_4 = 2$;
(c,b)	$\rho_5 = 2 + \frac{1}{2} = 2\frac{1}{2}$;
(a)	$\rho_6 = 3$.

Here the quantity ρ can be taken to mean the usual number. If $\rho = m$, an integer, then this corresponds to the case where $K(q)$ coincides with a linear subspace of $\{\dot{q}\}$ with dimension $m \leq n$.

Such an approach adopted for defining the degree of freedom of control u is guided by the fact that the quantity degree of freedom of control must reflect the property of a CDS to be controllable—the higher the degree of freedom of control the "more controllable" the CDS and, conversely, the lower the degree of freedom of control the "less controllable" the CDS, at least in the local sense, that is, in the neighbourhood of q. For example, if dim $f(q, U) = n$ and $\dot{O} \in$ ri $f(q, U)$ we can move the representative point of the CDS in all directions, that is, the vector $\dot{q}$ can be assigned an arbitrary direction in the space $\{\dot{q}\}$. It is pertinent to point our here that the degree of freedom of control is maximum and equals n.

It is interesting to observe that if u is a vector the degree of freedom of control u is, in general, not directly connected with its size. It is clear that, for example, a scalar control, that is, a one-dimensional control, can have different degrees of freedom depending on $f(q, U)$.

The Hamiltonian of CDS as a Support Function

We now define the Hamilton function, or the Hamiltonian, of a CDS by the formula

$$H(p, q) = \sup_{\dot{q} \in f(q, U)} p\dot{q}, \tag{1}$$

Where $p\dot{q}$ is the scalar product of the vectors p and $\dot{q}$ in R^n or, what is same, by

$$H(p, q) = \sup_{u \in U} P(p, q, u) = \sup_{u \in U} pf(q, u), \tag{2}$$

where $P(p, q, u) = pf(q, u)$. Clearly, $H(p, q)$ is nothing other than the support function of the convex set $f(q, U)$ [97]. If the upper bound in (1) or (2) is attained, we write

$$H(p, q) = \max_{\dot{q} \in f(qU)} p\dot{q} = \max_{u \in U} pf(q, u). \tag{3}$$

It is natural to assume that $H(0, q) = 0$. The function $H(p, q)$ is defined for all $p \in \{p\}$ and $q \in \{q\}$. For a given $p \neq 0$, at every fixed $q \in \{q\}$ the set $f(q, U)$ has an oriented support plane with normal p. The equation of this plane in $\{\dot{q}\}$ is of the form $p\dot{q} = H(p, q)$. The distance d between this plane and the point $\dot{O} \in \{\dot{q}\}$ is clearly

$$d = H\left(\frac{p}{|p|}, q\right). \tag{4}$$

From formulas (1) – (3) it is clear that for a fixed $q \in \{q\}$ the function $H(p, q)$ is a positive homogeneous function of degree one in p, that is,

$$H (\lambda\, p, q) = \lambda\, H (p, q) \qquad \forall\, \lambda \geq 0. \tag{5}$$

Differentiating (5) with respect to λ (p, q fixed) and setting $\lambda = 1$, we obtain the Euler's identity

$$p\, \frac{\partial H}{\partial p}\, (p, q) = H (p, q), \tag{6}$$

Where $\dfrac{\partial H}{\partial p}$ denotes, as usual, the gradient vector of H (p, q) with respect to p.

For a fixed q, the function H (p, q) is a convex function of p, that is, for any two vectors $p = p_1$ and $p = p_2$ the condition

$$H (\mu_1 p_1 + \mu_2 p_2, q) \leq \mu_1 H (p_1, q) + \mu_2 H (p_2, q), \tag{7}$$

where $\mu_1 \geq 0, \mu_2 \geq 0, \mu_1 + \mu_2 = 1$, is satisfied. In view of (5), the convexity condition (7) can be expressed in a simpler equivalent form

$$H (p_1 + p_2, q) \leq H (p_1, q) + H (p_2, q) \tag{8}$$

this follows immediately from (1) – (3).

The support function H (p, q) provides complete information regarding classes of CDSs which are equivalent in the generalized sense, the equivalence being expressed in terms of the set f (q, U) (Sec.3). Many important features and properties of a given CDS can be completely described in terms of the function H (p, q). In addition to the aforementioned properties of H (p, q) we list further properties determined by the nature of the convex set f (q , U).

1. In order that the set f (q , U) lies completely in a hyperplane having normal p, it is necessary and sufficient that [21]

$$H (p, q) = - H (- p, q) \quad \text{or} \quad H (p , q) + H (- p, q) = 0. \tag{9}$$

Here H (p, q) and H (– p, q) denote the distance between the support hyperplanes and the set f (q , U) in the directions $p \neq 0$ and –p. If (9) holds, the affine dimension of the set f (q , U) does not exceed n – 1. It then follows that f (q , U) is of affine dimension n, that is, is a domain containing (absolutely) interior points, if and only if there does not exist any $p \neq 0$ for which (9) holds. This property will be needed in Sec. 20 for identifying "singular manifolds" of the CDS.

2. If f (q , U) constitutes a domain in $\{ \dot{q} \}$ then a necessary and sufficient condition for the point $\dot{0}$ (q) to lie in the interior of this domain, that is for $\dot{0}$ (q) $\in$ int f (q, U), is that

$$H (p , q) > 0 \quad \text{for} \quad \forall\, p \in \{p\}, \quad p \neq 0, \tag{10}$$

or

$$\min_{|p| = 1} H (p , q) > 0. \tag{11}$$

if it is noted that H (p , q) is homogeneous in p. This property of H (p , q) will be used below (Sec.9) to identify the domain of unconstrained (free) motions of the CDS.

3. If relation (9) holds at the point q for two noncollinear vectors p_1 and p_2, the affine dimension of f (q , U) does not exceed n − 2, and so on.

4. If f (q , U) can be expressed in the form f (q , U) = f_1 (q , U) + a (q), where f_1 (q, U) is a convex set and a (q) a constant vector (for fixed q) or, in other words, if there is a translation, then

$$H (p , q) = H_1 (p , q) + p \, a (q), \tag{12}$$

where H_1 (p , q) is the support function of f_1 (q , U) .

5. The support function of the point a (q) is

$$H (p , q) = p \, a (q). \tag{13}$$

6. The support function H (p , q) of the unit sphere or, more precisely, of the ball $|p| \le r (q), r (q) > 0$, is

$$H (p , q) = |p| \, r (q). \tag{14}$$

7. When f (q , U) does not constitute a domain in $\{\dot{q}\}$ it is convenient to have necessary and sufficient conditions for the point $\dot{O}$ to lie in the relative interior of f (q ,U), that is, to have conditions under which $\dot{O}$ (q) ∈ ri f (q , U). This condition is of the form

$$H (p , q) > 0, \qquad \forall \, p \in \{ p \}, \qquad p \ne 0, \tag{15}$$

except for those p which satisfy (9).

Combining the above mentioned properties we can state a few more useful conditions in terms of H (p , q) guaranteeing the following assertions regarding the corresponding convex set.

8. f (q , U) lies completely in the n − m dimensional linear subspace $\{ \dot{q} \}$ of that is obtained by the intersection of hyperplanes determined by the nonzero vectors $p_1,...,p_m$ if and only if

$$H (- p_i , q) + H (p_i , q) = 0, \qquad i = 1, ..., m.$$

This is a generalization of property 1 stated above.

9. Let f (q , U) lie completely in the subspace determined by the vectors $p_1,...,p_m$. Then $\dot{O}$ ∈ $\{ \dot{q} \}$ lies in the relative interior of f (q , U) $\dot{O}$ ∈ ri f (q , U), if and only if

$$H (p , q) > 0$$

for all $p \neq 0$ except those which are linear combination of $p_1,...,p_m$.

We conclude this section with the following observation. In terms of H (p , q) the equation of a CDS can be written in the form

$$\dot{q} = \frac{\partial H}{\partial p} \ (p, q), \qquad (16)$$

where the parameter p can be treated as a new control which is *not subjected* to any condition apart from (16), and any value from { p } is admissible. Note that for a fixed $q \in \{q\}$, $\frac{\partial H}{\partial p} (p, q)$, regarded as a function of p, is a/positive homogeneous function of degree zero, that is,

$$\frac{\partial H}{\partial p} \ (\lambda p, q) = \frac{\partial H}{\partial p} \ (p, q) \qquad \forall \lambda > 0. \qquad (17)$$

Thus in the new Eq. (16) of the CDS, the velocity does not depend on the magnitude | p | of vector p but depends solely on its directions p / | p |. Hence, with no loss of generality, we can consider only those vectors (controls) p whose ends lie on a sphere of a fixed radius R, in particular, on the unit sphere, in the space { p }.

Eq. (16) can also be written in the form

$$\dot{q} = \frac{\partial H}{\partial p} \left(\frac{p}{|p|}, q \right). \qquad (18)$$

Unfortunately, an equation of the form (16) or (18) does not always possess a unique solution, for a fixed $q \in \{q\}$, in p (or p / | p |). Herein lies the main distinctive feature of many CDSs usually encountered in practice. Such a case is the one that generates the surface of integral funnel of inclusion (3.2) and occurs in extremely complex CDSs.

Eqs. (16) and (18) are solvable (that is, there exists an inverse function p of $\dot{q}$ for all $\dot{q}$ in a very particular, though quite common, case which can even be designated as classic from the point of view of mechanical systems) when the point Ȯ lies inside f (q , U) and f (q , U) is a domain in { $\dot{q}$ }.

It should be noted that no transformation of the original CDS to form (16) the quantity degf p of the new control, in the sense discussed in the preceding section, remains unchanged since the set of admissible velocities in the space { $\dot{q}$ } remains unchanged, namely f (q , U), that is, degf p = degf u.

Finally, a last remark. It is clear from (2) and (3) that for the computation of H (p , q) we need not actually construct convex hull of Φ (q) or f (q , U) if they are not already convex. Formulas (2) and (3) define H (p , q) automatically as a support function of the convex hull of Φ (q) or f (q , U). In this sense, we can speak of the support function H(p, q) of an arbitrary set Φ (q) or f (q , U) defined by (2) or (3). (We can also obtain as a support function of the convex hull of Φ (q) or f (q , U) the result will be the same.)

8

Types of Cones of Admissible Direction of CDS

All possible types of cones K(q) of admissible directions of a CDS have been shown in Table 8.1. The first row of this table shows that in the zero-dimensional subspace $L° \subset \{\dot{q}\}$ the cone K (q) can be of only one type. Namely, a single point coinciding with $\dot{O}$ (q). We denote this type by a_0^0 . The second row of the table shows that in the one-dimensional subspace $L° \subset \{\dot{q}\}$ the cone K (q) can be of two types. The first type coincides with L^1; this type is denoted by a_0^1. The second type of K (q) in L^1 coincides with a subspace of L^1 and is denoted by a_1^1 .

Next, the third row shows successively all types of two-dimensional cones K (q), that is, all possible cones for which $L^2 \subset \{\dot{q}\}$ serves as the minimal underlying subspace. The first square of this row contains all the cones coinciding with L^2. This type is designated as a_0^2 . The second square consists of those K (q) which coincide with a half-space of L^2; this type is denoted by a_1^2 . Finally, the third (last) square of this row depicts a cone of the type which has a pointed vertex and which completely lies in L^2. This type has been denoted by a_2^2 . These exhaust all types of *solid cones relative* to L^2. We thus find that these cones are of three types : a_0^2, a_1^2, a_2^2 If the first type a_0^2 is symbolically expressed by the intersection of two coordinate axes, then the transition $a_0^2 ---> a_1^2 ---> a_2^2$ corresponds to the successive removal ("chipping off") of the *negative semi-axes*.

Similarly, the fourth row of the table successively shows all possible types of *solid cones relative* to L^3. The first square of this row depicts a cone of the type coinciding with L^3; this is designated as a_0^3. The second square shows a cone of the type coinciding with a half-space of L^3; this has been denoted by a_1^3. This is followed by the cone of dihedral angle type represented by a_2^3. Finally, the last square again contains a tapered solid cone

28

relative to L^3; it is denoted by a_3^3. As in the preceding case, the transition $a_0^3 \dashrightarrow a_1^3 \dashrightarrow a_2^3 \dashrightarrow a_3^3$ may be regarded as the successive removal ("chipping off") of one semi-axis in each transition.

Table 8.1

Possible types a_r^m of cones K (q) of admissible velocity directions of an nth order CDS

	r = 0	r = 1	r = 2	r = 3	. . .	r = n
L^0 m = 0	a_0^0 $\dot{O}(q)$					
L^1 m = 1	a_0^1	a_1^1				
L^2 m = 2	a_0^2	a_1^2	a_2^2			
L^3 m = 3	a_0^3	a_1^3	a_2^3	a_3^3		
⋮	⋮	⋮	⋮	⋮	⋮	⋮
$L^n \equiv \{\dot{q}\}$ m = n	a_0^n	a_1^n	a_2^n	a_3^n	. . .	a_n^n

The fourth row, the fifth row etc. of the table contain successively types of solid cones relative to the subspaces L^4, L^5 etc.

The last row, the (n + 1) th row, of the table is associated with the velocity space of the given nth order CDS: that is, $L^n \equiv \{\dot{q}\}$ $T_q \equiv \{ q \}$. This row contains successively n + 1 types of cones: $a_0^n, a_1^n, ..., a_n^n$ The type a_0^n corresponds to the case where K (q) coincides with L^n. A cone of the a_0^n type contains n axes, that is, 2n semi-axes. The transition from each preceding type to the next amounts to removing one semi-axis. The last type, type a_n^n in the (n + 1) th row corresponds to the *tapered cone* K (q). Symbolically, n semi-axes have been removed from this cone.

Thus for an nth order CDS the general type of cone K (q) is denoted by a_r^m where m, m = 0, 1 ,..., n, is the dimension of the minimal underlying subspace $L^m \subset \{\dot{q}\}$ for K (q) and r, r = 0,1,...,m ($r \leq m$), is the number of the semi-axes removed ("chipped off") in the graphical representation of this type of K (q). The index (number) r can thus be called the "defect" in semi-axes. In the above classification for an nth order CDS the total number of types of cones K (q) equals

$$1 + 2 + 3 + ... + n + (n + 1) = \frac{(n + 1) (n + 2)}{2}.$$

Thus for a given L^m each row of the table contains successively all types of *solid cones relative* to L^m – from the first type a_0^m , coinciding with the entire L^m, to the last type a_m^m, the tapered cone. Each transition in the series of transitions in the row in question from a given square to the next one, $a_0^m \dashrightarrow a_1^m \dashrightarrow a_2^m \dashrightarrow ... \dashrightarrow a_m^m$, amounts to removing one semi-axis in the original representation of a_0^m , in the form of a system of coordinate axes with centre at the point $\dot{O}$ (q) .

For a given nth order CDS, it is interesting to find the type of the conjugate (dual) cone $\overline{K}$ (q) associated with the cone K (q) (see formula (3.8) of a given type a_r^m . The type of $\overline{K}$ (q) will be denoted by $\overline{a}_r^m$. It can be easily shown that

$$\overline{a}_r^m = a_r^{n-m+r} , \quad m = 0, 1 ,..., n ; \quad r = 0, 1 ,..., m. \tag{1}$$

The set of semi-axes for $\overline{a}_r^m$ supplements that for a_r^m to 2n, the maximum number of semi-axes.

Indeed, 2m – r semi-axes/for the type a_r^m added to 2 (n – m + r) – r semi-axes for the type $\overline{a}_r^m$ yield 2m – r + [2 (n – m + r) = 2n, that is, the number of semi-axes for a_0^n. It is also easy to check that $\overline{a}_r^n = a_r^r$. Instead of a_r^m and $\overline{a}_r^m$ we can straightway use the notation K_r^n (q) and $\overline{K}_r^m$ (q), respectively.

We shall formulate a criterion for the cone K(q) to be a definite type a_r^m at a given point q.

A criterion for determination of type of the cone. *For an nth order CDS, the* cone K (q) $\subset T_q \{ q \} \equiv \{\dot{q}\}$ is of type a_r^m at the fixed point q $\in$ {q} if the following conditions are satisfied : there exist exactly n – m + r linearly independent solutions

$$p = \pi_i (q) , \quad i = 1, 2..., n - m + r, \tag{2}$$

of the equation

$$H (p , q) = 0, \tag{3}$$

out of which exactly $n - m$ solutions $p = \pi_i (q)$, $i = 1 ,..., n - m$, also satisfy the system of equations

$$H (p , q) = 0, \tag{4}$$

$$H (-p , q) = 0. \tag{5}$$

The remaining r vectors of (2) will be denoted by $X_j (q)$, $j = 1, 2,...,r$.

It should be noted that the type a_r^m of the cone $K (q)$ was defined for a fixed $q \in \{ q \}$. Thus the type of a cone can change with q. Hence, generally speaking, the type of a cone, a_r^m, depends on the point q and is a function of this point: $a_r^m = a (m (q), r(q))$. In other words, the quantities m and r are functions of q, that is, $m = m (q)$, $r = r (q)$.

In view of this situation, one of the prime and fundamental problems of the theory of phase portrait of the CDS is the problem of *decomposition* of the entire/state space $\{ q \}$ of the given CDS (or of a given domain of this space) into disjoint sets (in particular, into domains or/into manifolds of smaller dimensions) over which the type of the cone $K (q)$ is saved (constant) and the value of a_r^m remains fixed.

The index m has a simple geometrical interpretation : it denotes the smallest dimension of the subspace $L^m \subset T_q \{ q \}$ which contains wholly the cone $K(q)$. This m will be called the *dimension* of the cone $K_r^m (q)$. It is clear that when $m = 0$ the cone degenerates into the point $\dot{O} (q)$ of the space $T_q \{ q \} \equiv \{ \dot{q} \}$. Such points will be referred to as the *points of absolute equilibrium* for the CDS. A condition for $K (q) \equiv \dot{O} (q)$ is clearly that $H (p , q) = 0$ for all p and a given q.

The cone $K (q)$ is called is *solid* cone if $m = n$. That is, solid cones are those cones $K (q)$ which are of type a_r^n, integer $r \leq n$.

The cone $K (q)$ is called *tapered* if $r = m$. That is, tapered cones are those cones which are of type a_m^m.

It follows from the above statement (formulas (2) - (5)) that the vectors q belonging to the cone $K (q)$ of a given type $a_r^m (q)$ are defined by a system of homogeneous linear equations.

$$\pi (q) \dot{q} = 0, \tag{6}$$

where the matrix $\pi (q)$ consists of $n - m$ rows $\pi_i (q)$, $i = 1, ... , n - m$, given by formulas (2), and is hence an $(n - m) \times n$ matrix with rank $\pi (q) = n - m$, and inequalities of the form

$$X(q)\dot{q} \leq 0, \tag{7}$$

where the matrix $X(q)$ consists of r rows $X_j(q)$, $j = 1, \ldots, r$, and is hence a r x n matrix with rank $X(q) = r$.

The system of Eqs. (6) can be treated as a set of $n - m$ equations defining $n - m$ linearly independent hyperplanes. And, in its turn, the set of inequalities (7) can be treated as a set of r inequalities defining r linearly independent half spaces of the space $T_q\{q\} = \{\dot{q}\}$ whose intersection contains $K(q)$. The hyperplanes determining these r half space constitute r linearly independent support hyperplanes of $K(q)$. The equation of the jth support hyperplane ($j = 1, \ldots, r$) is clearly of the form

$$X_j(q)\dot{q} = 0 \quad \text{or} \quad X_j(q)\,dq = 0. \tag{8}$$

We shall see later (Sec. 19) that the system (4), (5) plays a leading role in the investigation of a CDS. This system must be treated as system of Pfaffian equation (in terms of differentials)

$$\pi(q)\,dq = 0, \tag{9}$$

where $\pi(q)\,d\,q$ is the corresponding systems of I-forms or *Pfaffian forms* [95]. The system (9) defines, in particular, invariant manifolds of the given CDS.

We remark that the relations (7), (8) can be replaced by equivalent relations of exactly the same form but with the property that the matrix $X(q)$ is orthogonal to the matrix $\pi(q)$, that is,

$$\pi(q)X^T(q) = 0. \tag{10}$$

In other words, each row vector of $X(q)$ is orthogonal to every row vector of $\pi(q)$. The set of those points of $\{q\}$ where the cone $K(q)$ is of type a_r^m will be denoted by D_r^m.

Let $K(q)$ be a continuous field of cones, and let m_1 be the maximum value of m for the given CDS ($m \leq n$). Then the set $D_r^{m_1}$ of $\{q\}$ is an open region with dim $D_r^{m_1} = n$, because otherwise there would exist a point $q' \in D_r^{m_1}$ whose neighbourhood would contain points $q \in D_r^m$ where $m < m_1$. But then at least one of the rows of $\pi(q)$ would vanish which is impossible since, by hypothesis, the field $K(q)$ is continuous.

In general, if two sets $D_r^{m'}$ and $D_r^{m''}$ have a common boundary with distinct indices m' and m", m" > m', then the boundary points belong to $D_r^{m'}$, that is, to the set with smaller $m = m'$. In particular, if m_0 is the smallest value of m for the given CDS, then $D_r^{m_0}$ is a closed set. For instance, the (invariant) set of absolute equilibrium of the CDS is always closed.

If the field of cones $K(q)$ has "breaks" (is discontinuous), the set of discontinuities must be identified. The motion of a CDS on these sets requires special investigation.

Domain of Free Trajectories of CDS

Suppose that at every point q of a domain $D \subset \{q\}$ the convex set $f(q, U)$ also constitutes a domain in the space $\{\dot{q}\} = T_q\{q\}$ containing the point $\dot{O} \in \{\dot{q}\}$ in its interior. Then the cone $K(q)$ coincides with the entire space $\{\dot{q}\}$ that is, the cone is of type a_0^n. For the set $f(q, U)$ to constitute, indeed, a domain in $\{\dot{q}\}$ containing $\dot{O}$ for a given q, it is necessary and sufficient that the support function $H(p, q)$ of this set be strictly positive for all $p \neq 0$ (Sec. 7). Since the function $H(p, q)$ is homogeneous in p, this condition is equivalent to the condition

$$\min_{|p|=1} H(p, q) > 0 \qquad (1)$$

This inequality defines a domain D in the state space $\{q\}$. It is also clear that if such a domain D exists, then any two points q_0 and q_1 in the connected portion of this domain can be joined by an arbitrary trajectory which completely lies in D and which is the result of the action of an admissible control. It is natural to refer to such a domain D as the *domain of free trajectories* of the CDS. We observe in passing that every point q of this domain will be a point of local controllability.

Thus the coordinates of points q belonging to the domain of free trajectories D satisfy the inequality

$$\min_{|p|=1} H(p, q) \equiv \varphi(q) > 0. \qquad (2)$$

The set of boundary points q of D will satisfy the equation

$$\min_{|p|=1} H(p, q) \equiv \varphi(q) = 0. \qquad (3)$$

The equation to be satisfied by the boundary points of the domain of free trajectories can also be derived directly from the condition that $\dot{O} \in \partial f (q, U)$, that is, from the condition that the point $\dot{O}$ lies on the boundary of the domain of admissible velocities (on the the indicatrix). Assume that the equation of the boundary of the domain $f (q, U)$ (that is, of indicatrix) is of the form

$$\sigma (\dot{q}, q) = 0, \tag{4}$$

where q is a parameter. Then the required equation is of the form

$$\sigma (0, q) = 0. \tag{5}$$

Let us consider a simple example. Suppose that a CDS is governed by the equation

$$\dot{q}_1 = a q_2 + u_1, \quad q_2 = b q_1 + u_2, \quad U = \{ (u_1, u_2) \mid u_1^2 + u_2^2 \le R \}. \tag{6}$$

Here

$$P (p, q) = a p_1 q_2 + p_1 u_1 + b p_2 q_1 + p_2 u_2 ; \tag{7}$$

$$H (p, q) = \max_{u \in U} P (p, q, u) = a p_1 q_2 + b p_2 q_1 + R \sqrt{p_1^2 + p_2^2}; \tag{8}$$

$$\varphi (q) = \min_{|p| = 1} H (p, q) = -\sqrt{a^2 q_2^2 + b^2 q_1^2} + R = 0. \tag{9}$$

Thus the required equation of the boundary of the domain of free trajectories D is of the form

$$a^2 q_2^2 + b^2 q_1^2 = R^2, \tag{10}$$

which represents an ellipse.

The same result can be derived directly. We have from (6)

$$u_1 = \dot{q}_1 - a q_2, \quad u_2 = \dot{q}_2 - b q_1. \tag{11}$$

Noting the conditions prescribed by U, we obtain from (6)

$$\sigma (\dot{q}, q) \equiv (\dot{q}_1 - a q_2)^2 + (\dot{q}_2 - b q_1)^2 - R^2 = 0. \tag{12}$$

Setting $\dot{q} = 0$ in (12), we obtain the result coinciding with (10).

The Principle of Inclusion in the Event Space

We consider a simple principle which, though obvious, proves to be useful in solving various control problems. Let K (q (t), t) be the cone of admissible directions with vertex (q, t) constructed in the space (q, t).

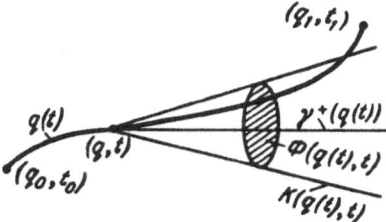

Fig. 10.1

The principle of Inclusion in the Event Space. For a curve $q = q(t)$, $t_0 \leq t \leq t_1$, lying in { q, t } to be an admissible integral curve, it is necessary and sufficient that the positive tangent ray $\gamma^+(q(t)$ to q (t) lies in the cone of admissible directions K (q (t), t), at almost all points (q (t), t) (see Fig. 10.1), that is,

$$\gamma^+(q(t)) \in K(q(t), t) \quad a \; \forall \, t \in [t_0, t_1].$$

The Boundary of an Integral Funnel of a Differential Inclusion

In the theory of differential inclusions of the form

$$\dot{q} \in \Phi(q, t), \tag{1}$$

where $\Phi(q, t)$ is, for every q and t, a given non-empty convex set in the event space { q, t), the set of points of absolutey continuous integral curves $q = q(t), t \geq t'$, of (1) originating from the point { q', t') is known as the *integral funnel with vertex* { q', t') for the differential inclusion (1). If integral curves are considered on a time segment (t', t_1), one can speak of a *segment of the integral funnel*. A segment of the integral funnel, with vertex { q', t') for the differential inclusion (1) will be denoted by V (q', t', t_1). We shall examine the case where the segment V (q', t', t_1) has a "rigid" lateral boundary in the form of a conoid with vertex { q', t' }.

The lateral boundary ∂ V (q', t', t_1) of a segment of the integral funnel V (q', t', t_1) will be called *rigid* in a neighbourhood (may be sufficiently small) of {q', t') if it is independent of t_1. In this case the notation V (q', t', t_1) can be replaced by V{q', t') and, similarly, ∂ V (q', t', t_1) by ∂ V (q', t'). Sufficient conditions for the existence of a rigid integral funnel are, in general, provided by the condition for the smoothness of the field of cones K { q', t') in the neighbourhood of { q', t') lies in an invariant manifold will be examined in Sec. 22. For the present, we assume that the conoid is solid in the space { q, t). This conoid has a plane base and forms a part of the plane $t = t_1$ (Fig. 11.1). In this case, we can speak of the *boundary* ∂ V as a lateral surface of the rigid integral funnel and state the flollowing proposition.

The boundary ∂ V of the integral funnel V (q', t', t_1) with vertex at the point { q', t'), for the differential inclusion (1) is defined by the parametric equation

$$q = q(t, p', p'_{n+1}', q', t'), \quad t' \leq t \leq t_1, \tag{2}$$

where (q, t') is a fixed point (the vertex of the funnel), and t and p' are parameters. The scalar parameter t varies on the segment $t' \leq t \leq t_1$, where t_1 is sufficiently small, and the nozero vector parameter $p' = (p'_1, ..., p'_n)$ and the scalar parameter p'_{n+1} vary, for fixed (q', t'), over the set of solutions of the equation

$$H(p', q', t') + p'_{n+1} = 0. \tag{3}$$

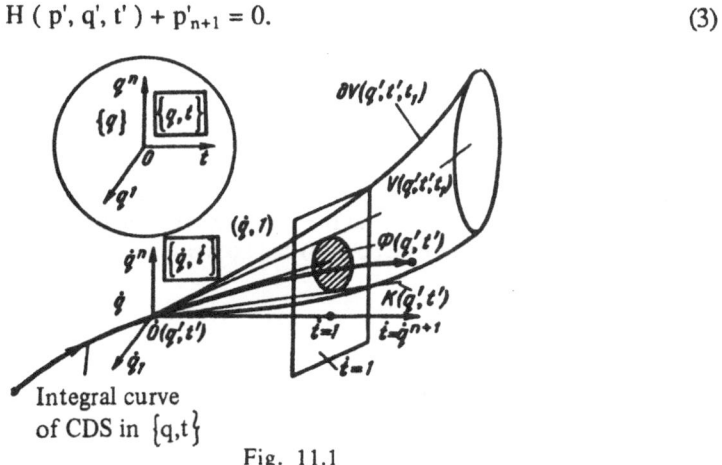

Integral curve
of CDS in $\{q,t\}$

Fig. 11.1

The function $H(p, q, t)$ in (3) is determined by the "Principle of upper bound"

$$H(p, q, t) = \underset{\dot{q} \in \Phi(q, t)}{\text{Sup}} p\dot{q}, \quad p = (p_1, ..., p_n), \tag{4}$$

and by the "maximum principle"

$$H(p, q, t) = \underset{\dot{q} \in \Phi(q, t)}{\max} p\dot{q} \tag{5}$$

if the upper bound in (4) is attained. In terms of differential equation of the CDS, $H(p, q, t)$ is determined by the formula

$$H(p, q, t) = \underset{u \in U(q, t)}{\text{Sup (max)}} pf(q, t, u). \tag{6}$$

Furthermore, the function (2) is determined as a solution of the Cauchy problem for the system of Hamilton's canonical equations

$$\dot{q} = \frac{\partial H}{\partial p}, \quad \dot{p} = -\frac{\partial H}{\partial q}, \quad \dot{p}_{n+1} = -\frac{\partial H}{\partial t}, \tag{7}$$

with the Hamiltonian (4)-(6), subject to the initial conditions

$$q\,(\,t'\,) = q'_0, \quad p\,(\,t'\,) = p', \quad p_{n+1}\,(\,t'\,) = p'_{n+1}, \tag{8}$$

where p' and p'_{n+1} vary over the set of solutions of Eq. (3).

On the other hand, in the space (q, t) the boundary of the integral funnel with vertex (q' t'), of the Hamilton-Jacobi equation

$$H\,(\frac{\partial z}{\partial q}, q, t\,) + \frac{\partial z}{\partial t} = 0. \tag{9}$$

This envelope constitutes a characteristic conoid for Eq. (9).

Thus the boundary of the integral funnel with vertex (q', t') coincides with the characteristic conoid of Eq. (9) with the same vertex. The equation of the boundary of the integral funnel (the characteristic conoid) can be written in the form

$$z\,(\,q, t\,) = 0, \tag{10}$$

where $z\,(\,q, t\,)$ is a solution of Eq. (9).

The above theorem clearly follows from the geometrical interpretation of the first order nonlinear partial different equation (9) in one unknown scalar function z (see, for example, [104]) O.

In conclusion, we remark that the function $p\,\dot q$ of variables p and $\dot q$ and the function pf (q, t, u) of variables p, q, t, u can be appropriately called the *Pontryagin* function (the *Pontryaginian*[1] and denoted by

$$P\,(\,p, \dot q\,) = p\,\dot q, P\,(\,p, q, t, u\,) = pf\,(\,q, t, u\,).$$

Eqs. (7) are nothing but the equations of extremals of the Pontryagin maximum principle that arise in connection with the extremal problem of the time functional for the CDS of types (1) and (2.1).

We note another important property of the integral funnels for convex connections, which can be termed the "embedding property" of integral funnels. In the space { q, t }, let V (q', t', t_1) be an integral funnel of the given inclusion $\dot q \in \Phi\,(\,q, t\,)$. Take a point (q", t") $\in \partial$ V (q', t', t_1). Then V (q", t", t_1) $\subset$ V (q', t', t_1). The funnel V (q', t', t_1), $t_1 > t'$, bounds partly the domain of reachability from the point (q', t') during the time $t' \le t \le t_1$. It should be remarked that in the event space { q, t } the domain of unconstrained motion is always absent (Sec. 9) because the given problem in its present formulation rules out reverse motion, that is, one can never reach the point (q", t") from (q', t') if t"< t'.

---

[1]This function was first investigated in inclusion with Pontryagin's maximum principle [88]. This was earlier noted in [110]

Relationship of the Boundary of an Integral Funnel with the Hamilton-Jacobi Equation

We condiser a nonlinear scalar partial differential equation of first order of the form

$$H (p, q, t) + p_{n+1} = 0, \tag{1}$$

where (q, t) are independent variables in the space $\{ q, t \}$, and $p = \dfrac{\partial z}{\partial q}$, $P_{n+1} = \dfrac{\partial z}{\partial t}$ and $z (q, t)$ is an unknown scalar function. A large number of remarkable works are devoted to this equation (see, for example, [62, 63, 101, 104]). A geometrical interpretation of this equation is as follows. We fix a point (q', t') in $\{ q, t \}$, and from this point draw a vector $(p, p_{n+1}) = (\partial z/\partial q, \partial z/\partial t)$. This vector is normal to the plane which passes through (q', t') and is tangent to the integral surface $z (q, t) = 0$; here $z (q, t)$ is a solution of Eq. (1). For a fixed (q', t'), the set of directions of vectors $(p, p_{n+1}) = (\partial z/\partial q, \partial z/\partial t)$ satisfying Eq. (1) constitues, in general, the surface (boundary) of a cone with vertex (q', t'). This cone will be denoted by $\overline{K} (q', t')$ and will be called, as is customary, the *normal (polar) cone*.

When the direction of the vector (p, p_{n+1}) runs through the boundary $\partial \overline{K} (q', t')$ the tangent plane thus created describes the surface (boundary) of another cone $K (q', t')$, which in the theory of Eq. (1) is referred to as the *tangential cone* or *the Monge's cone*. The surface $\partial K (q', t')$, is the envelope of the family of tangent planes to the surfaces $z (q', t') = 0$ through the point (q, t'), that is, $z (q', t') = 0$.

The way the cone $K(q', t')$ has been defined, it is a double-sheet cone. But since we wish to identify this cone with the cone of admissible velocity directions in the velocity space $\{\dot{q}, \dot{t}\}$ we retain only that sheet of the cone $K(q', t')$ which is directed along the positive t-axis. Henceforth, $K(q', t')$ will denote only this single sheet cone.

In the theory of equations of type (1), a decisive role is played by the notions of "characteristic curves" and "characteristic strips", and, in particular, by the notion of "characteristic conoid"; this conoid is tied to a certain point (q', t') at its vertex.

The central idea establishing a one-to-one correspondence between equations of form (1) on one hand, and between CDS (4.1) and differential inclusion of form (4.2) on the other lies in that one sheet of the Monge cone K (q', t') is indentified with the cone K (q', t') of admissible velocity directions for the differential inclusion (4.2) or the associated CDS. Thus in the theory of equation (1) the following problem arises : for the given (1), that is, for the given function H (p, q, t), construct the corresponding Monge cone. In order for us to be able to establish a link between Eq. (1) and a CDS, we must solve the *inverse problem* : for the given cone of admissible directions K (q, t) (the Monge cone), it is required to construct the differential equation (1), that is, to recover the function H (p, q, t).

Let us solve this problem. Assume that we are given a CDS

$$\dot{q} = f (q, t, u), \quad u \in U (q, t), \tag{2}$$

or the corresponding differential connection

$$\dot{q} \in \Phi (q, t) \equiv f (q, t, U (q, t)), \tag{3}$$

where the set Φ (q, t) is non-empty. For a fixed (q, t), we construct Φ (q, t) and the associated cone K (q, t) = conv (ray Φ (q, t)[1]. We then construct the conjugate cone $\overline{K}$ (q, t), and take a vector (p, p_{n+1}) $\in \partial\overline{K}$ (q, t). In view of the construction of K and $\overline{K}$ the relation

$$\underset{u \in U (q, t)}{\text{Sup}} \quad [p f (q, t, u) + p_{n+1}] = 0, \tag{4}$$

or, what is same, the relation

$$\underset{u \in U (q, t)}{\text{Sup}} \quad [p f (q, t, u)] + p_{n+1} = 0 \tag{5}$$

holds. The first term on the left side of (5) is the desired function

$$H (p, q, t) = \underset{u \in U (q, t)}{\text{Sup}} \quad p f (q, t, u) = \underset{\dot{q} \in \Phi (q, t)}{\text{Sup}} \quad p\dot{q}. \tag{6}$$

Indeed, it can be easily verified that the Monge cone for differential Eq. (1) with H in the form (6) is the desired cone of admissible directions K (q, t) for the systems (2) and (3).

To determine the characteristic conoid with vertex (q', t'), determined by the initial cone K (q', t'), or, in other words, to determine the surface of the integral funnel ∂ V (q', t', t), it is necessary to write Hamilton's canonical characteristic equations.

$$\dot{q} = \frac{\partial H}{\partial p} (p, q, t), \quad t \geq t', \tag{7}$$

---

[1] The symbol ray Φ (q, t) denotes the set of all those rays which originate from 0 (q, t) and have a non-empty intersection with Φ (q, t)   [97].

$$\dot{p} = -\frac{\partial H}{\partial q}(p, q, t), \quad t \geq t', \tag{8}$$

$$\dot{p}_{n+1} = -\frac{\partial H}{\partial t}(p, q, t), \quad t \geq t', \tag{9}$$

and for them solve the Cauchy problem with the initial conditions

$$q(t') = q', \tag{10}$$

$$(p(t'), p_{n+1}(t')) = (p', p'_{n+1}) \in \partial \overline{K}(q', t'). \tag{11}$$

The condition (11) is equivalent to the condition

$$H(p', q', t') + p'_{n+1} = 0 \tag{12}$$

which is what is stated in the theorem of the preceding section. It can be early seen that the vector $(p(t), p_{n+1}(t))$ is normal to the surface of the funnel. It should also be remarked that the conditions

$$p\, dq + p_{n+1}\, dt = 0 \tag{13}$$

is satisfied on the surface of the integral funnel.

In particular, if the CDS in question is autonomous, that is, if the right hand sides of the equation of motion and of the differential inclusion are independent of t, then $\partial H/\partial t = 0$ and $\dot{p}_{n+1} = 0$. Consequently, $p_{n+1}(t) = $ const. in the autonomous case.

In the theory of relativity [115], the cone $K(q, t)$ is known as the *light cone*; it corresponds to the constant velocity of light $(c = 1)$. The surface of the integral funnel can be called the *light conoid*; this corresponds to the case of variable velocity of light which is propagated in an inhomogeneous non-isotropic medium (for instance, in the gravitational field; see also Sec. 30).

The Principle of Inclusion for an Autonomous CDS in the State Space Integration of CDS

For an autonomous CDS the principle of inclusion in the space { q, t } can be stated in the phase space { q }. We take a point q in { q } and for this point construct the cone K (q̇) in { q̇ }. This cone consists of all rays originating from q along directions of admissible vectors $\dot{q} \in \Phi (q)$.

The Principle of Inclusion for Autonomous System in the Phase Space. Let $\bar{q}$ be a segment of a non-selfintersecting piecewise smooth curve in the phase space { q }, and let q_0 be the initial point and q_1 the final point of this segment. For the given curve to be an admissible trajectory, it is necessary that the positive ray of the tangent $\gamma^+ (q)$ to the curve $\bar{q}$ belongs to the cone of admissible directions K (q) for almost all points of $\bar{q}$, that is,

$$\gamma^+ (q) \in K (q) \ a \ \forall \ q \in \bar{q} \tag{1}$$

That the condition is necessary is obvious. What is more, under fairly broad additional assumption this becomes sufficient as well. Indeed, let $\bar{q}$ be a non-selfintersecting curve in { q } with initial point q and final point q_1, and let $\gamma^+ (q) \in K (q) \ a \ \forall \ q \in \bar{q}$. In accordance with the construction of the cone K (q) of admissible directions, the ray $\gamma^+ (q)$ has a non-empty intersection with $\Phi (q)$. We take a point (q) of this intersection. This point determines a definite branch of the differential inclusion $\dot{q} \in \Phi (q)$, namely,

$$\dot{q} = \phi (q). \tag{2}$$

In fact, $\phi (q)$ determines a control $u \in U$ at the point q.
We assume that the *time integral*

$$t = \int_{q_s}^{q_1} \frac{|dq|}{|\varphi(q)|} \qquad (3)$$

is defined and assumes a finite value. Then it is clear that the additional assumption regarding finiteness of integral (3), incorporated in the inclusion principle, renders this principle sufficient as well for the curve $\bar{q}$ to be an admissible trajectory. Usually condition (3) implies that during the motion along a chosen curve $\bar{q}$ there should be no point of equilibrium of differential Eq. (2) from which an exit under the action of an admissible control would be impossible. Such a point is known as a *point of absolute equilibrium*. For instance, (0,0) is a point of absolute equilibrium for the bilinear system of the form $\dot{q} = A q + u B q$.

Though the inclusion principle stated here is trivial in nature, it nevertheless enables us to easily check whether or not a given trajectory is admissible.

Here it is pertinent to draw the following analogue. If we have an UDS governed by the equation

$$\dot{q} = f(q), \quad q \in \{q\}, \qquad (4)$$

then this equation can be treated as the *field of directions* in the space $\{q\}$. The *problem of integration* of this equation consists in finding curves (trajectories) in $\{q\}$ such that the direction of tangents to them coincides at every point with the direction of the field at the point.

If we have a CDS governed by the equation

$$\dot{q} = f(q, u), \quad u \in U, q \in \{q\}, \qquad (5)$$

or by the connection

$$\dot{q} = \Phi(q), \quad q \in \{q\}, \qquad (6)$$

then conditions (5), (6) can be regarded as conditions determining the *field of cones* $K(q)$ in $\{q\}$. In analogy with the above statement, we can define the *problem of integration of CDS* (5) *or inclusion* (6) as a problem of seeking curves (trajectories) in $\{q\}$ such that the direction of tangents to them at every point q belongs to the cone $K(q)$ of field of cones at the point. In other words, the problem of integration of CDS (5) or of inclusion (6) lies in determining curves (trajectories) in $\{q\}$ which satisfy the inclusion principle in the state space.

The geometrical method of integrating UDS (4) (it is a consequence of the problem of integration itself) is known as the *method of isoclines* (see [119]).

For the integration of (5) or (6) a similar geometrical method can also be suggested. This method consists in choosing in $\{q\}$, say, in the particular case n = 2, in the plane or on any other two-dimensional manifold, a sufficiently dense web of fixed points q near which the cones $K(q)$ are drawn. In a diagram, the size of $K(q)$ must naturally be small—of the order of size of elementary cells conditioned by the denseness of the chosen web of fixed points q. The external side of each such cone can be hatched so as to clearly see the sheet of the cone that determines the admissible directions. If we have such a diagram, then we can draw the trajectory satisfying the inclusion principle.

14

Boundary of the Trajectory Funnel of CDS

If the given CDS is autonomous, that is, if it is governed by the equation

$$\dot{q} = f(q, u), \quad u \in U(q), \quad t \geq 0, \tag{1}$$

or by the corresponding inclusion

$$\dot{q} \in \Phi(q), \quad t \geq 0, \tag{2}$$

where f, Φ and U do not depend on t explicitly, we can speak of a *trajectory funnel* of the given CDS. By a *segment of the trajectory funnel* V (q', T) of CDS (1) or (2), we mean the set of all admissible trajectories of CDS (1), originating from the point q', along which the representative point of the CDS moves during the time T > O. The point q' is called the *vertex of the funnel* V (q', T).

However, in contrast to the integral funnel V (q', t', t_1) in the event space { q, t }, which never completely fills up (since always $\dot{t} = 1 > 0$) the neighbourhood of its vertex { { q', t' }, the trajectory funnel can fill up this neighbourhood. In other words, if the convex cone K (q') of admissible velocity directions with vertex q' coincides with the tangent space T_q, { q } at this point, then trajectory funnel completely fills up the neighbourhood of q' in the state space { q }. In this case, the neighbourhood of q' is the domain of unconstrained (free) trajectories, discussed earlier in Sec. 9.

An interesting case, though more complex that encountered frequently in practice, is the one where the convex cone K (q') does not coincide with the tangent space T_q, { q } in which it has been constructed. Then, generally speaking, the segment of the trajectory funnel V (q', T), at least for sufficiently small T > O, also does not fill up the neighbourhood of q' in the state space { q }[1]. We assume that the trajectory funnel

---

[1]For the time being, we exclude from our discussion the singular points and the locally reachable points.

44

V (q', T) is solid. This implies that no invariant manifold passes through q' (Secs. 19, 22).

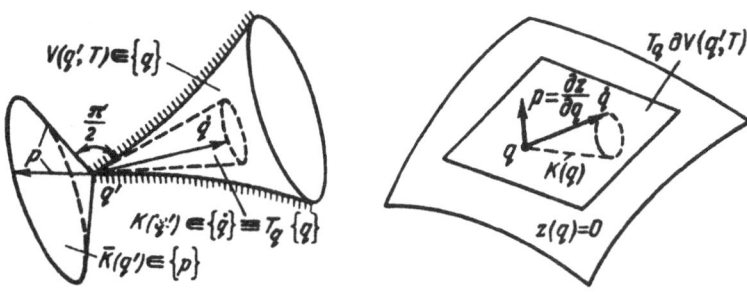

Fig. 14.1   Fig. 14.2

If the trajectory funnel V (q', T) is *rigid* (that is, if its lateral surface does not change with T in a sufficiently small interval [0,∈], ∈ > 0), then the funnel V (q', T) will have locally, at least, a rigid lateral surface (boundary) denoted by ∂V (q'). The notion of rigid trajectory funnel is similar to that of a rigid integral funnel examined in Sec.II. Generally, if the field of cones K (q) is sufficiently smooth in the neighbourhood of q and all the cones are tapered, then ∂ V (q') exists.

Such a funnel has the form of a conoid with vertex (point of tapering) at the point q' ; for this conoid we can identify the *lateral boundary surface* (or the *lateral boundary* of the funnel) and its base (Fig. 14.1). In this case, the lateral boundary surface and the base of the conoid together constitute the complete boundary of the segment of the trajectory funnel V (q', T). Clearly, the base of the conoid is composed of those points q ∈ { q } which are reachable from the point q' in the given time T > 0 by means of admissible controls *in the optimal speed process.*

We recall (Sec. 11) that in the space { q, t } the base of the integral funnel is always flat planar; it lies in the hyperplane $t = t_1$ (or in t = T when t' = 0). But the trajectory funnel is, in general, not a planner hypersurface; this surface can be referred to as the *Bellman surface.*

From the point of view of the theory of control, the trajectory funnel V (q', T) constitutes the *domain of reachability* from the initial point q' in time T.

What is important is that we shall be interested not in the base of the rigid funnel but in its lateral boundary V (q'). Indeed, the main problem in the present section is to derive the equation of the lateral boundary of the rigid funnel V (q').

As in the case of the integral funnel (Sec. 11), the derivation of the equation to the rigid boundary of the trajectory funnel can be based on the identification of the cone K (q) with the Monge cone of the corresponding nonlinear scalar partial differential equation of first order. The same equation can also be derived directly as illustrated in Fig. 14.2

In fact, suppose that the equation of the boundary of the funnel is of the form

$$z (q) = 0.$$ (3)

On ∂V (q'), we take an arbitrary point q at which there exists a normal p = $\partial z/\partial q$ to ∂V (q'). Then the tangent plane $T_q \partial V$ (q') to the surface ∂V (q') at the point q is a support plane and tangential to the cone K (q) (Fig. 14.2). Consequently

$$H (p, q) = \quad \underset{\dot{q} \in \Phi (q)}{\text{Sup (max)}} p\dot{q} = \quad \underset{u \in U (q)}{\text{Sup (max)}} \quad pf (q,u) = 0, \quad (4)$$

where p = $\partial z/\partial q$ at q. We thus obtain the desired equation for z (q) of Eq. (3) :

$$H (\frac{\partial z}{\partial q}, q) = 0. \tag{5}$$

This also yields the parametric equation of the boundary of the trajectory funnel. Indeed, let

$$\dot{q} = \frac{\partial H}{\partial p}, \ \dot{p} = - \frac{\partial H}{\partial q}, \ t > 0, \tag{6}$$

be the Hamilton's canonical equations corresponding to (5), and let the functions

$$q = q (t, q', p'), \quad t \geq O, \tag{7}$$

$$p = p (t, q', p'), \quad t \geq O, \tag{8}$$

be a solution of the system (6) subject to the initial conditions

$$q (o, q', p') = q' \tag{9}$$

$$p (o, q', p') = p' \tag{10}$$

where the point q' (the vertex of the funnel) is fixed and the vector p' runs through the conjugate cone K (q'), that is, p' runs through the set of solutions of the equation

$$H (p', q') = 0. \tag{11}$$

Then Eq. (7) represents the parametric equation of the boundary of the trajectory funnel with vertex q', the parameters being t and p'.

The solution (8) gives us the motion of the normal vector to ∂V (q') along the "generator" of the trajectory (7). Using a familiar terminology [104], the funnel V (q') can be called the "characteristic conoid" of Eq. (5) or of system (6).

On the other hand, if we have the complete integral of Eq. (5)

$$z = z (q, c_1 ,..., c_n), \tag{12}$$

where $c_1,..., c_n$ are arbitrary constants, then the boundary ∂V (q', T) of the funnel can be determined as the enveloping surface of the family of surfaces

$$z (q, c_1,...,c_n) = 0 \qquad (13)$$

passing through q', in the space {q}.

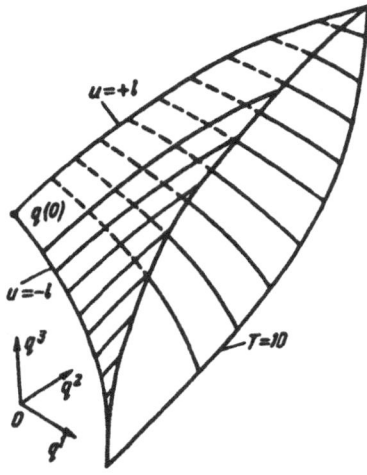

Fig. 14.3

It should be observed that to each nonsingular point $q \in \partial V (q')$ there corresponds a normal p to $\partial V (q')$. In this way, we obtain a mapping of the surface $\partial V (q')$ in/to the phase space { p, q } of the Hamilton's system (6). The image constitutes an n-dimensional surface known as a "conical surface". It is easy to see that on this surface the constraint of the form pdq vanishes identically [9, pp 326-332]. Fig. 14.3 depicts a segment of isometric trajectory funnel for a third-order CDS :

$$\dot{q}^1 = q^2, \quad \dot{q}^2 = q^3, \quad \dot{q}^3 = u, \quad |u| \leq l = 20, \, q' = (10, 10, 10), \quad T = 10.$$

15

Geometrical Construction of the Boundary of Trajectory Funnel

Consider a CDS governed by the equation

$$\dot{q} = f(q, u), \quad u \in U(q), \tag{1}$$

or by the corresponding differential inclusion

$$\dot{q} \in f(q, U), \tag{2}$$

where $f(q, U)$ is a convex set in $T_q\{q\}$. Take a point $q' \in \{q\}$ and with this point as vertex draw the cone $K\{q'\}$ of admissible velocity directions for the given CDS (1) or (2). Then the boundary (surface) of the trajectory funnel $V(q, T)$ with vertex q' can be constructed, for $T > O$, (this surface is assumed to exist in the form of a solid conoid) by approximating it by a "broken" surface depending on a small parameter $\tau = T/N$, where N is a sufficiently large quantity. These approximating surfaces converge, as $\tau \rightarrow 0$, to the desired surface $V(q', T)$.

With no loss of generality, we shall assume that the lengths (moduli) of all the vectors $\dot{q}$ are unity, that is, $|\dot{q}| = 1$ (Sec. 5). From the point q' draw all possible vectors, $\tau\dot{q}$ with $\tau > O$ a small quantity, lying on the surface of the cone $K(q,)$. All these vectors are of equal length τ. Geometrically, the set of endpoints of these vectors constitutes a curve l on the surface $\partial K(q')$. On the other hand, the endpoints of vectors $\tau\dot{q}$ are assumed to be points of the state space $\{q\}$ of CDS (1) or (2).

We draw cones $K(q'')$ at all points q'' of l, treated as a curve in $\{q\}$ (Fig. 15.1). This yields a family of cones $K(q'')$ which depdends on the parameter $q'' \in l \subset \{q\}$. We construct the external enveloping surface Π' for the family $K(q'')$. Since Π' is an invelope of cones, it is a ruled surface. From each point $q'' \in l'$, we draw a vector which $\tau\dot{q}, |\dot{q}| = 1$, lies simultaneously on the surface $\partial V(q'')$ of the cone and on the ruled envelope Π'_o. The endpoints of vectors $\tau\dot{q}$ again constitute a curve l' in $\{q\}$.

48

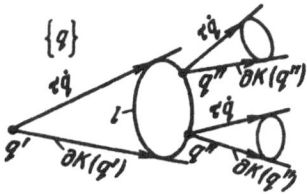

Fig. 15.1

With points of l' as vertices, we construct a new family of cones K (q' ' ') for which the point q ' ' ' ∈ l' serves as a parameter. We then find the external enveloping (ruled) surface Π" for this family of cones K (q'''), q''' ∈ l' ⊂ { q }. This process is continued.

This process results in a broken striped surface, each strip of which is a ruled surface. The width of each strip is of the order τ. The surface thus obtained is the approximating surface of the disired surface of the trajectory funnel V (q', T). When τ --> 0. the approximating surfaces, in general, tends to V (q', T). It should be remarked that the construction of the strips of the approximating surface for the trajectory funnel need not be carried out under the assumption that the vectors are normalized, that is, it is unnecessary to assume that $\mid \dot{q} \mid = 1$. The construction can be carried out taking the natural length of the vector, that is, taking the length determined by the indicatrix of the CDS (1) or (2). It is clear that in a similar fashion the boundaries of the integral funnel can be drawn in the event space { q, t }.

16

The Euler-Lagrange Equation for the Boundary of Trajectory Funnel

In Sec. 11, for admissible trajectories lying on the boundary of the trajectroy funnel V (q') in { q } we derived the Hamilton's canonical equations

$$\dot{q} = \frac{\partial H}{\partial p}\,(p, q), \quad \dot{p} = -\frac{\partial H}{\partial q}\,(p, q), \tag{1}$$

with the initial conditions

$$q (o) = q', p \quad (o) = p', \tag{2}$$

where q' is the fixed vertex of V (q') and p' is the vector parameter whose components run through the set of solutions of the equation.

$$H (p', q') = 0. \tag{3}$$

Geometrically, Eq. (3) implies that the vector p' varies over the surface of the conjugate cone $\overline{K}$ (q), that is, $p' \in \partial \overline{K}$ (q'). The equation of this cone is furnished by Eq. (3). For a CDS governed by the equation

$$\dot{q} = f (q, u), u \in U (q), \tag{4}$$

or by the inclusion

$$\dot{q} \in f (q, U), \tag{5}$$

the Hamiltonian H (p, q) was defined by the formula

$$H (p, q) = \text{Sup (max) pf} (q, u) = \text{Sup (max) p } \dot{q}. \tag{6}$$
$$u \in U (q) \qquad \dot{q} \in \Phi (q)$$

50

This approach employed for describing the boundaries of trajectory funnels can be termed as the "Hamiltonian formalism". However, one can adopt a different approach, dual to this one, based directly on the field of cones K (q).

The idea to be employed in the derivation of the Lagrange's equation for admissible trajectories constituting the boundary of a rigid trajectory funnel can be based on the geometric method, discussed in Sec. 15, for constructing boundaries of trajectory funnels. This is, in fact, an analogue of the Euler's method of broken lines used in the calculus of variations for the derivation of extremal equations.

On the other hand, the desired Euler's equations can be derived from the canonical Eqs. (1)-(3). For a fixed $q \in \{ q \}$, let $L (\dot{q}, q) = 0$ be the equation of the cone K (q) of admissible velocity directions for the CDS (4) or (5). To keep the discussion simple, it is assumed that the function $L (\dot{q}, q)$ is differentiable a required number of times and K (q) is a tapered cone having dimension n. Having regard to appropriate solvability and differentiability conditions, we can write the following equations

$$p = \frac{\partial L}{\partial \dot{q}} \tag{7}$$

$$\dot{q} = \frac{\partial H}{\partial p} \tag{8}$$

whence

$$\dot{p} = \frac{d}{dt} \left(\frac{\partial L}{\partial \dot{q}} \right) \tag{9}$$

Furthermore, it is easy to see that

$$\frac{\partial H}{\partial q} = - \frac{\partial L}{\partial q} . \tag{10}$$

Substituting (9) and (10) into the second equation of system (1), we obtain the desired Euler-Lagrange ordinary diffeential equations

$$\frac{d}{dt} \left(\frac{\partial L}{\partial \dot{q}} \right) - \frac{\partial L}{\partial q} = 0 \tag{11}$$

for the boundary of the trajectory funnel V (q').

In contrast to the 2n equations of first order in system (1), Eq. (11) is a system of n equations each of which is a differential equation of second order. The initial conditions for q of (11) are same as in (2), that is,

$$q (0) = q'. \tag{12}$$

The second initial condition concerning of $\dot{q}$ (11), necessary for the system (11) to have a unique solution, is of the form

$$\dot{q}(0) = \dot{q}' \; ; \tag{13}$$

here the components of the vector $\dot{q}'$ play the role of a parameter which runs through all solutions of the equation of the cone $K(q')$:

$$L(\dot{q}', q') = 0. \tag{14}$$

The geometrical significance of the initial conditions (14) is that the initial vector $\dot{q}'$ varies over entire surface of the cone $K(q')$.

Thus if we move over from the Hamiltonian formalism to the Lagrangian formalism for describing the surface of the funnel $V(q')$, we have integrate Eq. (11) instead of Eq. (1).

Let us consider a simple example illustrating the application of the Lagrange equations for determining the form of the trajectory funnel. Let the CDS be governed by the vector equation

$$\dot{q} = a + u, \quad q \in R^n, \quad u \in U \subset \{\dot{q}\} \equiv R^n, \quad |u| \le r, r > 0. \; = 0. \tag{15}$$

where the vector a and the scalar r depend, in general, on q and $|a| \ge r$. Thus the set of endpoints of the admissible velocities of CDS (15) constitutes in $\{\dot{q}\}$ a ball

$$|\dot{q} - a| \le r \tag{16}$$

of radius r with centre a. Consequently, the indicatrix of (15) is given by the equation

$$\sigma(\dot{q}, q) |\dot{q} - a| - r = 0. \tag{17}$$

It can be easily seen that in the same space the corresponding cone $K(q)$ of admissible velocity directions is a circular cone given by the equation

$$L(\dot{q}, q) \equiv a\dot{q} - \sqrt{a^2 - r^2} |\dot{q}| = 0, \tag{18}$$

where the function $L \dot{q}, q = a\dot{q} - \sqrt{a^2 - r^2} |\dot{q}|$ is the Lagrangian of the CDS (15).

If a and r do not depend on q, then the Lagrange Eq. (11) assumes the form

$$\frac{d}{dt}\left(\frac{\partial L}{\partial \dot{q}}\right) - \frac{\partial L}{\partial q} \equiv \frac{d}{dt}\left(a - \sqrt{a^2 - r^2}\, \frac{\dot{q}}{|\dot{q}|}\right) = 0, \tag{19}$$

hence

$$a - \sqrt{a^2 - r^2}\, \frac{\dot{q}}{|\dot{q}|} = c \tag{20}$$

in which c is an arbitrary constant vector. From (20) we obtain

$$\frac{\dot{q}}{|\dot{q}|} = \frac{a - c}{\sqrt{a^2 - r^2}} = c_1,$$ (21)

where c_1 is an arbitrary constant vector.

Eq. (21) shows that the vector $\dot{q}$ retains a constant direction along the characteristic curve. Consequently, the characteristic curve in the state space { q } of CDS (15) is a straight line. In view of the initial condition L ($\dot{q}'$, q') = 0, we find that the characteristic conoid of the given CDS or, what is same, the trajectory funnel with vertex at an arbitrary point q' of the state space is a circular cone (a particular case of a conoid) with vertex q' ∈ { q }. This cone is described by the equation

$$a (q - q') - \sqrt{a^2 - r^2} \, | q - q' | = 0.$$ (22)

Eq. (22) describes a family of trajectory funnels of CDS (15) in its state space, the role of a parameter of this family being played by q'.

It should be observed that if | a | < r the cone K (q) does not exist in the proper sense of this term. In this case, K (q) coincides with the entire space { $\dot{q}$ }, and the entire state space { q } becomes the domain of unconstrained (free) trajectories (Sec. 9).

Hatched Boundaries of Trajectory Funnels Manifold of Reverse Hatching (MRH)

In the present section, we introduce the notion "hatching of boundaries of trajectory funnels". The meaning of this term will become clear from the following discussion. We shall consider the case of rigid funnels.

We assume that the boundary $\partial V (q', T)$ is, at least locally in the neighbourhood of the point q', a two-sided hypersurface in $\{ q \}$ and that the cone $K (q)$, for $q \in \partial V (q', T)$, does not lie completely in the tangent plane to $\partial V (q', T)$ at the point q. The side of the boundary $\partial V (q', T)$ to which $K (q)$ is adjoined for $q \in \partial V (q', T)$ will be referred to as the *internal side* of $\partial V (q', T)$ and the other side, naturally, as the *external side*.

We perform hatching on the external side of $\partial V (q', T)$. The boundary $\partial V (q', T)$ together with the hatchng will be known as the *hatched boundary* of the trajectory funnel.

The points $q \in \partial V (q', T)$ for which the cone $K (q)$ lies *wholly* in the tangent plane $T_q \partial V (q', T)$ will be called the *singular point* of $\partial V (q', T)$. The set of singular points on the given boundary $\partial V (q', T)$ will be known as the *singular manifold of the boundary* of the given trajectory funnel.

The singular manifold on $\partial V (q', T)$ is a submanifold of the boundary $\partial V (q', T)$ itself. It can have any dimension—from zero (a single point) to the full dimension. As we shall see, in many cases singular manifolds of the boundaries of trajectory funnels play a significant role in the construction of a phase portrait of the given CDS and reflect important properties of the CDS itself (its controllability, for instance) ; this becomes absolutely clear if one considers the case of a second-order CDS.

The significance of the hatching carried out on $\partial V (q', T)$ lies in that any trajectory which intersects $\partial V (q', T)$ at a point q, where hatching is defined, in moving from unhatched side to the hatched side (that is, intersects the boundary of the funnel when one moves from the internal side to the external side) happens to be inadmissible. This result is also clear from the fact that in this case the principle of inclusion is violated (Sec. 13). On the other hand, any admissible trajectory necessarily intersects the boundary of the trajectory funnel only in accordance with the hatching (that is, from the external side to the internal side), and this again agrees with the principle of connection.

It should be noted that in exactly the same fashion one can define on the basis of the principle of inclusion the notion of the "hatched surface (boundary) of the integral funnel" in the event space { q, t } : the motion of the CDS (admissible integral curves) can take place only in accordance with hatching but not vice-versa.

Thus in accordance with the meaning just discussed, hatching can also be performed on singular parts of the boundary of trajectory (integral) funnels.[1]

Let us examine another important aspect of the notion of hatched boundary of trajectory funnel. It can happen that the boundary (hypersurface) of the given trajectory funnel extends continuously beyond certain limits but after this limit the hatching of the continuously extending surface is *reversed*. Such a continuous extension of the boundary of the trajectory funnel can be determined as a continuous extension of the corresponding solution z (q) of the Hamilton-Jacobi equation or as an extension of solutions of the Hamilton's characteristic equations (Sec. 14).

It is pertinent to observe that hatching can be defined on any part of the surface given by the equation z (q) − 0, where z (q) satisfies the Hamilton-Jacobi equation for the given CDS, that is, z (q) satisfies the equation

$$H \left(\frac{\partial z}{\partial q}, q \right) - 0. \tag{1}$$

We now introduce another notion which is vital for the theory of phase portrait of a CDS. A manifold of dimension n − 2 on the hypersurface of the trajectory funnel which separates (at least locally) parts of this surface with hatching in opposite directions will be referred to as the *manifold of reverse hatching on the given boundary of the trajectory funnel*. Likewise, a separating manifold on the surface z (q) = 0, where z (q) satisfies Eq. (1), will be referred to as the *manifold of reverse hatching on the given surface* z (q) = 0.

The aggregate of all such manifolds for all the trajectory funnels of a given CDS an all the surfaces z (q) = 0, where z (q) satisfies Eq. (1), will be known as the *manifold of reverse hatching* of the CDS or, in short, MRH. This manifold, if it exists and if the CDS has no invariant manifold (Sec. 19), is, in general, of dimension n − 1, that is, is a hypersurface in the state space { q } of the given CDS. The notion of an MRH enables us to determine the limits for extending the proper trajectory of the funnel V (q '), justifying thereby the term "funnel"[2]. It can be stated that the *boundary* ∂V (q') of the given funnel V (q') with vertex at the point q' stretches only up to the point where it meets the MRH. This definition of the boundary of the trajectory funnel, which is no more local, is illustrated by a mechanical or hydromechanical diagram (Fig. 17.1). If the vertex q' of the funnel V (q') is assumed to be a source of fluid flowing into this funnel, then fluid particles by moving along admissible trajectories can wet the external side of the funnel (say, in the neighbourhood of q') only by spilling through the edge (brim) of the funnel which lies in the MRH.

---

[1] We shall not here carry out hatching on the invariant mainfold of the CDS with respect to the underlying space.

[2] As noted earlier, for a rigid funnel we use the shorter notation V (q') instead of V (q', T).

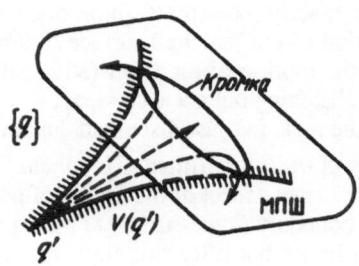

Fig. 17.1

The notion of an MRH plays a significant role in the investigation of a CDS. Therefore it is desirable to have constructive conditions for its determination. This question is dealt with in Sec. 18.

Singular Manifolds of CDS in State Space

In the preceding section we introduced singular manifolds of reverse hatching on the boundary of a rigid trajectory funnel $V(q')$. It is clear that for the existence of a singular manifold of reverse hatching it is necessary that there exist points $q \in \partial V(q')$ such that $\dim K(q) < n$.

Assume that the given CDS is governed by the equation

$$\dot{q} = f(q, u), \quad u \in U(q), \quad q \in \{q\}, \quad t > 0, \tag{1}$$

or the connection

$$\dot{q} \in \Phi(q), \quad q \in \{q\}, \quad t > 0, \tag{2}$$

where

$$\Phi(q) = f(q, U(q)). \tag{3}$$

It is known (Sec. 14) that if $z(q) = 0$ is the equation of the boundary $\partial V(q')$ of the trajectory funnel $V(q')$, then $z(q)$ satisfies the Hamilton-Jacobi equation

$$H\left(\frac{\partial z}{\partial q}, q\right) = 0. \tag{4}$$

On the other hand, if $q \in \partial V(q')$ is a singular point, then $K(q)$ and, hence, $\Phi(q)$ lie completely in the tangent plane to $\partial V(q')$ at q. Let $p = \partial z/\partial q$ be normal to $\partial V(q')$ at q. Then the "thickness" of the set $\Phi(q)$ in the direction of $p = \partial z/\partial q$ q is zero. Consequently, by formula (7.9), we have

$$H\left(\frac{\partial z}{\partial q}, q\right) + H\left(-\frac{\partial z}{\partial q}, q\right) = 0,$$

which, by virtue of (4), yields a system of equations

$$H\left(\frac{\partial z}{\partial q}, q\right) = 0, \quad H\left(-\frac{\partial z}{\partial q}, q\right) = 0, \tag{5}$$

for points q of S. It is interesting to treat (5) as a system of, in general, nonlinear, partial differential equations in the entire state space { q } :

$$H(p, q) = 0, \quad H(-p, q) = 0, p = \frac{\partial z}{\partial q}. \tag{6}$$

The system (6) plays an important role in the investigation of a CDS. It will be demonstrated in Sec. 19 that this system, in particualr, determines invariant manifolds of the given CDS. Compatibility conditions for (6) determine manifolds which we shall call singular manifolds of the CDS and denote by S. Equations for S will be written in the form

$$s(q) = 0. \tag{7}$$

The theory regarding system (6) is well developed (see, for instance, [104]). This system can be reduced to a linear system, and the compatibility conditions for this system are derived by completing it to a full system. Such a completion can be achieved by a systematic application of the Poisson's brackets.

As an illustration, consider the linear CDS

$$\dot{q} = A q + bu, \quad q \in R^n, \quad u \in R^1, |u| \le 1, \quad 1 > 0. \tag{8}$$

For system (8) we have

$$H(p, q) = \max_{|u| \le 1} p(Aq + bu) = pAq + 1|pb|. \tag{9}$$

And system (6) becomes

$$pq + 1|pb| = 0, \quad -pAq + 1|pb| = 0, \tag{10}$$

whence we obtain a system of linear partial differential euqations

$$\frac{\partial z}{\partial q} A q = 0, \quad \frac{\partial z}{\partial q} b = 0. \tag{11}$$

The full system associated with (11) is of the form

$$\frac{\partial z}{\partial q} A q = 0, \frac{\partial z}{\partial q} b = 0, \quad \frac{\partial z}{\partial q} A b = 0, \quad ..., \frac{\partial z}{\partial q} A^{n-2} b = 0. \tag{12}$$

It contains n equations. A necessary and sufficient condition for system (12) to have a nontrivial solution is that

$$s(q) = |Aq, b, Ab, ..., A^{n-2}b| = 0. \tag{13}$$

If the determinant (13) vanishes identically (and his happens if the vectors b, Ab ,..., $A^{n-2}b$ are linearly dependent), then all points of the state space { q } are singular, that is, S coincides with { q }.

If the vectors b, Ab, ..., $A^{n-2}b$ are linearly independent, then Eq. (13) defines a manifold of dimension n–1 in {q} which is the desired singular manifold S. In the present case, the singular manifold is an (n-1)-dimensional plane through the origin $O \in \{q\}$, that is, S is a linear subspace of the state space. It should be remarked that if an MRH (Sec. 17) for CDS (8) exists, it is defined by Eq. (13). It will be found later (Sec. 19) that if the plane (13) happens to be simultaneously an integral surface for system (1.1) (this will be the case if $A^{n-1}b$ is linearly dependent on b, Ab, ..., $A^{n-2}b$), then this plane turns out to be an isolated invariant manifold for CDS (8).

It should be observed that if the parameters A and b of (8) are such that the determinant (13) vanishes identically, then this implies the existence of an invariant manifold of the given CDS (Sec. 19).

As a second example, consider the problem of finding an MRH for the bilinear CDS

$$\dot{q} = Aq + uBq, \quad q \in R^n, \quad u \in R^1, \quad |u| \le 1, \quad 1 > 0. \tag{14}$$

It can be easily seen that in this case the system, equivalent to (7), is of the form

$$pAq = 0, \quad pBq = 0, \quad p[AB]q = 0, ..., \tag{15}$$

where $p = \partial z/\partial q$ and [A,B] is the commutator of matrices A and B. The relevant condition for the system (15) to have a nontrivial solution is again the vanishing of an nth order determinant :

$$s(q) \equiv |Aq, Bq, [A, B]q, ..., | = 0. \tag{16}$$

Similar conditions are also obtained for a singular manifold S for the general CDS governed by an equation of the form

$$\dot{q} = f_0(q) + u f_1(q), \quad q \in R^n, \quad u \in U \subset R^1, \quad |u| \le 1. \tag{17}$$

For this system, Eqs. (7) become

$$pf_0(q) = 0$$
$$pf_1(q) = 0 \tag{18}$$

and the singular manifold S for this CDS is defined by the vanishing of an nth order determinant :

$$s(q) \equiv |f_0(q), f_1(q), [f_0(q), f_1(q), ..., | = 0, \tag{19}$$

where $[\,f_0(\,q\,),f_1(\,q\,)\,]$ are the Poisson's brackets of the vector functions $f_0(\,q\,)$ and $f_1(\,q\,)$.

Thus if the conditions (16), (19) are not satisfied identically, these equations are satisfied by the coordinates of points of S for these CDSs. The reduction of (16) or (19) into an identity implies the presence of a nontrivial solution to the system (15) or (18), respectively, and, hence, implies the presence of invariant manifolds, discussed below (Sec. 19).

To conclude the present section, we remark the following. The problem of deriving conditions for singular manifolds in the sense of the theory of optimal control, investigated in [34], is, as a matter of fact, close to the problem of finding the singular manifold S. The optimal control is understood in the sense of minimization of the terminal functional $\varphi(\,q\,(T)$ over the trajectories of CDS (1). Recall that in the sense of optimal control a singular manifold is a manifold in $\{\,q\,\}$ on which the function $P(\,p,q,u\,)=pf(\,q,u\,)$ attains its maximum over u for all points $u\in\omega$, where ω is a subset of U containing at least two distinct elements.

It should, however, be remarked that the problem of finding a manifold in the present case is somewhat narrower than the analogous problem in the theory of optimal control (understood as in [34]). The singular manifold in the theory of optimal control includes the singular manifold S of a CDS. The matter is that (in the terminology of boundary of trajectory funnel V (q') adopted in the present book) a point $q\in V(\,q'\,)$ will be singular in the sense of optimal control theory not only when $K(\,q\,)\subset T_q\,\partial V(\,q'\,)$ but also when some *flat* face of the cone K (q) lies in the tangent plane $T_q\,\partial V(\,q'\,)$ (the cone K (q) may not lie wholly in $T_q\,\partial V(\,q'\,)$. But q will not be a singular point of the boundary $\partial V(\,q'\,)$ of funnel in the abovementioned sense : in this case hatching is uniquely defined in view of the fact that the cone is directed inwards of the funnel, and, hence, at q there are admissible velocity directions $\dot q$ which force the representative point of CDS of move inside the funnel at a nonzero angle to the plane $T_q\,\partial V(q')$.

It is clear that conditions (7) can also be employed for finding singular manifolds in the sense of optimal control theory if for the cone of admissible directions not the entire cone K (q) is taken but only its flat face (or faces if the face is not unique) which also constitutes a cone $K_\omega(\,q\,)$ corresponding to the set $\omega\in U$ mentioned above.

Finally, it should be remarked that a condition of type (19) requires further justification and needs to be made precise. The description of the singular manifold S is vital for the investigation of structure of the phase portriat in the neighbourhood of S and that of controllability questions of the CDS. However, as far as the author is aware, from this angle the significance of "compatibility" conditions of system (6) on manifolds of smaller dimension or at individual points of $\{\,q\,\}$ has not been sufficiently clarified, in the mathematics literature.

Invariant Manifold of CDS

An *invariant manifold* M for the CDS

$$\dot{q} = f(q, u), \quad u \in U(q) \quad \text{or} \quad \dot{q} \in \Phi(q) = f(q, U), \tag{1}$$

is a manifold in the state space { q } with the following property. If, at a certain instant t = t', the representative point q (t') of CDS (1) lies inside this manifold, that is, if q' = q (t') ∈ ri M ⊂ { q }, then q (t) ∈ M for all t ⊂ [t', t' + ε] for, at least, a certain ε > 0. Since we have here a sufficiently general system of form (1), we are restrained to talk only of local definitions which are true, in general, only in sufficiently small neighbourhoods of the point q (t') and time t'.

The invariant manifold/M of dimension m will be denoted by M^m. It is clear that an invariant manifold M of a CDS is simultaneously a singular manifold S, that is, M ⊂ S. However, the converse truns out to be valid not always, though sometimes it does; an example is the linear CDS discussed in Sec. 20.

It is also plain that for a manifold M, given by the equation F (q) = 0, to be invariant for CDS (1) it is necessary and sufficient that $\frac{\partial F}{\partial q} f(q, u) = 0$ identically for q ∈ M and u ∈ U (q).

It is clear that the entire state space { q } is an invariant manifold for CDS (1). And, generally, if the CDS is defined right from the beginning on a manifold M (say, controllable motion of a point is studied on a shpere, torus etc.), then M is the state space and, by definition, an invariant manifold.

It is interesting to obtain constructive conditions (equations) that describe M. It is also clear that, in general, M may not be unique and there can be whole families of invariant manifolds M depending on some scalar parameters. Thus, for example, if Eq. (1) has integrals, independent of the control parameter u, of the form

$$F^k(q) = C_k = \text{const}, \; k = 1, ..., m \,, \; 1 \leq m \leq n, \tag{2}$$

that is,

$$\frac{\partial F^{\kappa}(q)}{\partial q} f(q, u) = 0, \qquad k = 1, ..., m, \tag{3}$$

for any $u \in U(q)$, then the manifold determined by the (finite) system of Eqs. (2) is an invariant manifold M. This manifold depends on the m parameters $c_1, ..., c_m$ and is of dimension $n - m$. A unique manifold from the family (2) can be determined by means of some initial condition, that is, by prescribing the representative point at a certain instant t':

$$F_k(q(t')) = 0, \qquad k = 1, ..., m. \tag{3'}$$

It is plain that the minimum dimension of the invariant manifold M passing through the point $q \in M$ must be dim $k(q)$. Thus there arises the question of obtaining conditions under which invariant manifolds of CDS (1) can be found. In other words, there arises the problem of determining functions $F^{\kappa}(q)$ satisfying (3).

Using the same arguments as in the derivation of conditions in connection with the singular manifold S (Sec. 18; an invariant manifold is a particular case of S), we find that the desired functions $F^{\kappa}(q)$ must be functionally independent integrals of the system of equations

$$H\left(\frac{\partial z}{\partial q}, q\right) = 0, \qquad H\left(-\frac{\partial z}{\partial q}, q\right) = 0, \tag{4}$$

In this way, the question of finding invariant manifolds reduces to that of finding all integrals, regular as well as singular, of system (4). We need not, naturally, deal with the question of integration of (4), since a vast literature is devoted to this question (see [39, 62, 95, 104], for instance). We only remark that if the integral, to which there corresponds a singular manifold of dimension n-1, exists, it can be derived by "algebraic" means as an algebraic condition for compatibility of (4) with respect to components of the vector $\partial f/\partial q$. Having found such a manifold, it remains only to check whether or not it is an integral manifold, and this can be easily done. For instance, in Sec. 18 such a manifold was defined, by formulae (13), (16) and (19) for CDSs of various types. This manifold is an integral manifold provided the function s(q) in these relations satisfies system (18.7).

Thus if for the given CDS there exists an invariant manifold M, we can consider the question of finding boundaries of trajectory funnels as well as the question of finding the domain of unconstrained (free) motions on M. In particular, such problems arise automatically when a CDS is defined right from the beginning on a manifold (on a sphere, a torus etc., for instance). The question of finding the domain of unconstrained (free) motions on M will be dealt with in Sec. 20.

Of course, if the invariant manifold in, say, $\{q\} = R^n$ is defined by Eqs. (2), these equations enable us to eliminate m variables from Eqs. (1) by lowering the order of the CDS up to $n - m$. However, this approach may turn out to be inapplicable. Therefore it is desirable to have equations of boundaries of trajectory funnels on M directly in terms of CDS (1) and Eqs. (2). To this question is devoted Sec. 21.

As an illustration of a CDS which has invariant manifolds, consider the following bilinear CDS (Bloch's equations [30]) in $\{q\} = R^3$:

$$\dot{q}^1 = -a\,q^2, \quad \dot{q}^2 = -a\,q^1 + u\,q^3, \quad \dot{q}^3 = -u\,q^2, \quad u \in U \subset R, \tag{5}$$

where a is a constant quantity and us is a scalar control. This system is a particular case of the system of the form

$$\dot{q} = f_0(q) + u\,f_1(q). \tag{6}$$

For a CDS of form (6), the system (4) becomes

$$\frac{\partial F}{\partial q}\,f_0(q) = 0, \qquad \frac{\partial F}{\partial q}\,f_1(q) = 0. \tag{7}$$

The same system can also be obtained directly from (3). A completion of this system leads to the system

$$\frac{\partial F}{\partial q}\,f_0(q) = 0, \qquad \frac{\partial F}{\partial q}\,f_1(q) = 0, \qquad \frac{\partial F}{\partial q}\,[\,f_0(q), f_1(q)\,] = 0, \tag{8}$$

where $[\,f_0(q), f_1(q)\,]$ are the Poisson's brackets.
   If (6) is the bilinear CDS

$$\dot{q} = A\,q + u\,B\,q \tag{9}$$

where A and B are constant square matrices, then system (8) assumes the form

$$\frac{\partial F}{\partial q}\,A\,q = 0, \qquad \frac{\partial F}{\partial q}\,B\,q = 0, \qquad \frac{\partial F}{\partial q}\,[\,A, B\,]\,q = 0, \tag{10}$$

where $[\,A, B\,] = AB - BA$ is the commutator of A and B. For the CDS (5) in question

$$A = \begin{bmatrix} 0 & -a & 0 \\ a & 0 & 0 \\ 0 & 0 & 0 \end{bmatrix} \quad B = \begin{bmatrix} 0 & 0 & 0 \\ 0 & 0 & 1 \\ 0 & -1 & 0 \end{bmatrix}$$

and consequently

$$[A, B] = \begin{bmatrix} 0 & 0 & -a \\ 0 & 0 & 0 \\ a & 0 & 0 \end{bmatrix} \tag{11}$$

$$[\,AB\,]\,q = (-aq^3, \quad 0, \quad aq^1)^T. \tag{12}$$

   The compatibility condition for system (10) or, what is same, the condition for MRH is of the form

$$s(q) \equiv |\, Aq, Bq\, [\, A, B\,]\, q\, | = 0. \tag{13}$$

A computation of $s(q)$ shows that

$$s(q) = \begin{vmatrix} -aq^2 & 0 & -aq^3 \\ aq^1 & q^3 & 0 \\ 0 & -q^2 & aq^1 \end{vmatrix} \equiv 0 \quad \forall\, q \in R^3 \tag{14}$$

that is, Eq. (13) is satisfied identically. This implies that all points of the state space of the CDS are singular points (in the sense of Sec. 18) and that this CDS has an invariant manifold because system (10) has nontrivial integrals. And, hence, CDS (5) has nontrivial integrals, independent of control u, of form (2).

The identity (14) thus shows that one of the equations in system (10) is "redundant" since it is algebraically deducible from the remaining two equations of the system—the columns Aq, Bq and [A, B] q are linearly dependent for any $q \in R^3$. But at the same time any two columns are linearly independent. Thus out of the three equations in (10), we can retain any two; for example, the equations

$$\frac{\partial F}{\partial q} A q = 0, \qquad \frac{\partial F}{\partial q} B q = 0. \tag{15}$$

These two equations constitute a full system.

Since the number of equations $m = 2 < n = 3$, this system has, in accordance with the general theory of such a system [51], an integral basis having dimension $n - m - 3 = 2 = 1$, that is, has one fundamental integral (true, here one has to verify a posteriori that $\frac{\partial F}{\partial q}(q)$ does not vanish identically in any subdomain of $\{ q \} = R^3$). Indeed, employing the standard technique for integrating the given system, we see that the fundamental solution is of the form

$$F(q) = \frac{1}{2}[(q^1)^2 + (q^2)^2 + (q^3)^2], \tag{16}$$

and $\partial F/\partial q = (q^1, q^2, q^3)^T \neq 0$ in any subdomain of $\{ q \} = R^3$. Thus CDS (5) does have invariant manifolds subject to the presence of integral (16). It is plain that the invariant manifold is a sphere

$$\frac{1}{2}[(q^1)^2 + (q^2)^2 + (q^3)^2], = \frac{1}{2}c^2 \tag{17}$$

of an arbitrary radius $c > 0$ having centre at the origin of the state space $\{ q \} = R^3$.

It is interesting to observe that in the present example the invariant manifolds (concentric and, hence, disjoint spheres) stratify the state space $\{ q \} = R^3$.

In the above example of the bilinear CDS in R^3, the matrices A and B were skew-symmetric. We shall see right now that the properties of the bilinear CDS mentioned here are also true for a bilinear CDS in R^n with arbitrary skew-symmetric matrices.

Recall that in the general case a square matrix $A = (a_j^i)$, $i, j = 1, ..., n$, with real number elements a_j^i is said to be skew-symmetric iff

$$A = A^T = 0 \quad \text{or} \quad \text{iff} \quad a_j^i + a_i^j = 0; \quad i, j = 1, ..., n.$$

It is plain that the quantities a_j^i situated on the principal diagonal of a skew-symmetric matrix vanish, while two entires which are symmetrically situated with respect to the principal diagonal are equal in magnitude but opposite in sign.

Theorem 1. The commutator $[A, B] = AB - BA$ of two skew-symmetric matrices A and B is a skew-symmetric matrix.

Proof. We have

$$[A, B]^T = (A, B)^T - (B, A)^T = B^T A^T - A^T B^T = BA - AB = - [A, B],$$

as required.

Theorem 2. The quadratic form $q^T Aq$ with a skew-symmetric matrix A vanishes identically. That is, if A is a skew-symmetric matrix, then $q^T Aq = 0$ for any $q \in R^n$.

Proof. We have

$$\frac{\partial}{\partial q} (q^T A q) = (A + A^T) q = 0$$

which shows that $q^T Aq$ is a constant independent of q. But, clearly, $q^T Aq = 0$ when $q = 0$. Hence $q^T Aq = 0$ as required.

Theorem 3. The nth order determinant of the form $| A_1 q ... A_n q |$, where A_k, $k = 1, ..., n$ are skew-symmetric matrices, vanishes identically. That is,

$$\Delta (q) \equiv | A_1 q ... A_n q | = 0 \qquad \forall q.$$

Proof. We assume the contrary, that is, we assume that there is vector $q* \neq 0$ such that $\Delta (q*) \neq 0$. Then q_* has a component $q*^k \neq 0$. We multiply this component by the kth row of $\Delta (q)$. By our hypothesis, the resulting determinant is also nonvanishing. Furthermore, this new determinant remains unchanged (and hence nonvanishing) if to its kth row are added all the rows multiplied respectively by $q_*^1, ..., q_*^{k-1}, q_*^{k+1}, ..., q_*^n$. On the other hand, this results in a new determinant whose kth row contains the elements $q^T A_1 q, ..., q^T A_n q$ respectively. But, by Theorem 2, all these elements vanish because they are values of the quadratic form with skew-symmetric matrices $A_1, ..., A_n$. Thus the resulting determinant vanishes, contrary to our assumption. It is plain that $\Delta(q) = O$ for $q = 0$. Hence $\Delta(q) = 0$ for all q.

Putting together Theorems 1, 2 and 3, we have :

Theorem 4. Let A and B be skew-symmetric matrices, and let $c_1, ..., c_m$ be matrices obtained by all possible, including multiple, commutation of A and B. Then the determinant consisting of any columns of the form Aq, Bq, $c_1 q, ..., c_m q$ vanishes identically for all q.

Theorem 3 enables us to deduce useful properties of the bilinear CDS of the form

$$\dot{q} = Aq + u_1 B_1 q + ... + u_m B_m q, \tag{18}$$

where $A, B_1, ..., B_m$ are skew-symmetric matrices. One comes across such bilinear systems in practical problems. For instance, the CDS governed by the Bloch's equations (19.5) is of this forms.

For CDS (18), the system of Eqs. (19.4) (that is, the system of equations for the singular manifold S) assumes the form

$$\frac{\partial F}{\partial q} A\dot{q} = 0, \frac{\partial F}{\partial q} B_1 q = 0, ..., \frac{\partial F}{\partial q} B_m q = 0. \tag{19}$$

Applying Theorem 3 to (19), we obtain the important results that this system is always complete. This is an immediate consequence of Theorem 3 since, as can be easily verified, Poisson's brackets for any pair of the vectors Aq, $B_1 q, ..., B_m q$ vanish identically for all q. Hence system (19) is not only complete but involutory as well, although a priori it is not in the Jacobian form. By the way, to the latter form it can always be reduced.

A CDS of form (18) has always an integral of the form

$$F_1(q) \equiv (q)^2 = c_1, \qquad c_1 \geq 0. \tag{20}$$

This can be easily demonstrated by multiplying both sides of Eq. (1) by q and applying Theorem 3. Thus, for this CDS the representative point q always moves on a sphere of a definite radius with centre at the origin. It is easy to see that the system of equations

$$\dot{p} = -(pA + u_1 p B_1 + ... + u_s p B_s), \tag{21}$$

which is adjoint of system (18), has the integral

$$F_2(p) \equiv (p)^2 = c_2, \qquad c_2 \geq 0, \tag{22}$$

whence we find that for a CDS of the form (18) the conjugate impulse p is of constant length. What is more, the Hamilton's system of 2n equations (18), (21) with the Hamiltonian

$$H(p, q) = p A q + u_1 p B_1 q + ... + u_m p B q \tag{59 A}$$

has another integral

$$F_3 (p, q) \equiv pq = c. \tag{23}$$

That is, the scalar product of the radius vector of the representative point q and the impulse p is a constant quantity. This can also be shown directly. Indeed, multiplying (18) by p and (21) by q and adding the resulting equations, we have

$$p \dot{q} + p \dot{q} = \frac{d}{dt} (p q) = 0, \tag{24}$$

and this leads to the new integral (23). By the way, (23) can also be obtained by Jacobi's theorem. This theorem states that if $F_1 (p, q)$ and $F_2 (p, q)$ are two integrals of the Hamilton's system (18), (21), then the Poisson's bracket

$$F_3 = [F_1, F_2] = \frac{\partial F_1}{\partial q} \cdot \frac{\partial F_2}{\partial p} - \frac{\partial F_1}{\partial p} \cdot \frac{\partial F_2}{\partial q} \tag{25}$$

is also an integral of the system (18), (21). It should only be remarked that, unfortunately, this new integral may turn out to be either trivial or an algebraic corollary of the integrals F_1 and F_2. However, this is not the case for the system (18), (21) in question, and we have a new nontrivial integral

$$F_3 = [(q)^2, (p)^2] = 2 p q = c,$$

coinciding with (23).

It is easy to see that integral (23) remains valid for an arbitrary CDS whose Hamiltonian H(p, q) is a homogeneous function of degree one in p and q. It should also be noted that for the CDS (18) the new Poisson's brackets $[F_1, F_3]$ and $[F_2, F_3]$ do not provide additional integrals independent of F_1, F_2, F_3. Straightforward computation demonstrates the validity of this statement.

To conclude the present section, we remark that depending on the form of the CDS the question of finding its invariant manifolds can be investigated as a "dual" of the technique put forward in the present section [39, p. 125]. Namely, instead of investigating systems of partial differential equations, we can investigate the corresponding Pfaffian equations (forms) defined by formulas (8.9) [95].

Singular and Invariant Manifolds of Linear CDS

Let us examine in detail the procedure for obtaining singular and invariant manifolds in the case of a linear CDS. Suppose that the given CDS is linear. Eq. (19.7) yields the system of equations

$$\frac{\partial F}{\partial q} A q = 0 ; \qquad \frac{\partial F}{\partial q} B = 0. \tag{1}$$

In accordance with the general theory of such systems [104], we complete it to a full system by forming all possible Poisson's brackets $L_{ij} = [\, L_i, L_j \,]$ starting with

$$L_1 = \frac{\partial (\cdot)}{\partial q} A q \qquad \text{and} \qquad L_2 = \frac{\partial (\cdot)}{\partial q} b. \tag{2}$$

We then obtain a system of homogeneous linear partial differential equations of the form

$$\frac{\partial F}{\partial q} A q = 0, \qquad \frac{\partial F}{\partial q} b = 0, \qquad \frac{\partial F}{\partial q} Ab = 0, \qquad ..., \qquad \frac{\partial F}{\partial q} A^{n-2} b = 0, \tag{3}$$

in the unknown function $F(q)$.

A condition necessary for this system to be compatible (that is, a condition under which there exists a nontrivial solution $F(q) \neq 0$ is, obviously, purely algebraic in nature. This condition requires that system (3), regarded as a system of linear homogeneous algebraic equations in the components of the vector $\partial F/\partial q$, has a nontrivial solution. As is well known, this condition is in the form of the vanishing of the determinant $s(q)$:

$$s(q) \equiv |\, Aq \quad b \quad A b \quad ... \quad A^{n-2} b \,| = 0. \tag{4}$$

The problem now reduces to the investigation of s (q), that is, of determinant (4) regarded as a function of q $\in$ R^n. Here there are at least three principal cases which we shall discuss below in detail one by one.

Case 1. The function s (q) does not vanish identically and does not satisfy the system of Eqs. (2) with

$$F (q) \equiv s (q). \tag{5}$$

In this case, the singular manifold S, defined by Eq. (4) is an (n – 1)-dimensional hypersurface passing through the origin of the state space { q } = R^n of CDS (1). In other words, S is an (n – 1)-dimensional linear subspace of { q }. But, since S is not an integral manifold of system (2), S is an MRH for the given CDS (1).

We shall demonstrate later in the present section that this case occurs for CDS (1) if the vectors b, Ab, ..., A^{n-2}b, A^{n-1}b are linearly independent (that is, if system (1) is controllable in the Kalman's sense [49]).

The linear CDS (1) has thus no invariant manifold, and MRH is determined by the hyperplane (4) if the linear CDS (1) is controllable in the Kalman's sense.

Case 2. The function s (q) does not vanish identically but satisfies the system (2) with F (q) $\equiv$ s (q). In this case, the singular manifold S, given by Eq. (4) is again an (n – 1)-dimensional subspace in { q } = R^n. But S is now an invariant manifold of CDS (1), and it has no other invariant manifold, apart from (4).

We shall show below in the present section that this case occurs if the vectors b, Ab, ..., A^{n-2}b are linearly independent but the vector A^{n-1}b is linearly dependent on them.

It should be remarked that, in general, S can be, besides being an invariant manifold, simultaneously an MRH. But in view of the fact that an MRH is simultaneously an invariant manifold, there is no admissible trajectory of CDS (1) which would intersect MRH. (This is the situation in Example 5, Sec. 35. In Fig. 35.6, the q^2 -axis is the invariant manifold S and no admissible trajectory intersects it. However, the boundaries of trajectory funnel *continuously* extend from one side of the q^2 -axis to the other side and the reversal of hatching takes place on the q^2 -axis itself).

On the other hand, we can easily cite an example of a CDS, illustrating Case 2, where S, given by Eq. (4), is invariant but not an MRH in the abovementioned sense.

$$\overset{.}{q}^1 = - q^2 , \overset{.}{q}^2 = u, \quad | u | \leq 1, \quad 1 > 0. \tag{6}$$

Indeed, it can be easily seen that for CDS (6) the invariant manifold S coincides with the q^2 -axis. However, the boundaries of trajectory funnels approach the q^2 -axis from both left and right sides only asymptotically, going off to infinity in upward and downward directions along the q^2 -axis. Hence in this case no two boundaries of trajectory funnels situated on either side of the q^2 -axis, that is, on either side of S, are continuously joined. This explains why in the present case the invariant manifold S is not an MRH.

Case 3. Suppose that s (q), that is, determinant (4), vanishes identically. This happens if the vectors b, Ab, ..., A^{n-2} b are linearly dependent. In this case, the system (2) has linear fundamental integrals, and the corresponding invariant manifolds are linear manifolds stratifying the entire state space of CDS (1). Naturally, this case corresponds to uncontrollability of CDS (1) in the sense of Kalman. But in this case, uncontrollability of CDS (1) is, if one may use the phrase, "deeper" in contrast to Case 2 - the state space is stratified into linear invariant manifolds of dimension n-2 or of lower dimension.

We now return to the question raised in connection with Case 1.

Theorem. In order that the function s (q), given by (4), does not vanish identically and does not satisfy the system of Eqs. (2), it is necessary and sufficient that the vectors   b, ab, ..., A^{n-2} b, A^{n-1} b are linearly independent (that is, CDS (1) is controllable in the sense of Kalman).

Proof (of Sufficiency). Assume that the vectors b, ab, ..., A^{n-1} b are linearly independent. Then, obviously,

$$\frac{\partial s(q)}{\partial q} A^{n-2} b = | A^{n-1} b \quad b \quad Ab ... A^{n-2} b | \neq 0, \tag{7}$$

which shows that the function F (q) ≡ s (q) does not satisfy the last equation of system (3) for any q, and more so for any point of manifold (4). Thus the given F (q) does not satisfy (3) and, hence, system (2). This also implies that s (q) does not vanish identically for otherwise it would satisfy systems (2) and (3), which is not the case.

Necessity. Assume that s (q) ≠ 0 and that s (q) does not satisfy system of Eqs. (2), (3) at points q for which (4) is true. We shall show that then the vectors b, Ab, ..., A^{n-1}b are linearly independent. We assume the contrary, that is, assume that b, Ab,..., A^{n-1}b are linearly dependent. If the first n − 1 vectors b, Ab, ..., A^{n-2}b were linearly dependent, then s (q) ≡ O, by virtue of (4), contradicting the condition that s (q) ≠ O. Thus b, Ab, ..., A^{n-2}b are, indeed, linearly independent. But it may still happen that A^{n-1}b is a linear combination of b, Ab, ..., A^{n-2}b . We shall now show that that is impossible under the assumptions made. We again assume the contrary, and let A^{n-1}b be a linear combination of b, Ab, ..., A^{n-2}b . That is, let there exist quantities μ_0, μ_1 , ..., μ_{n-2} such that

$$A^{n-1}b = \mu_0 b + \mu_1 Ab + ... + \mu_{n-2} A^{n-2}b. \tag{8}$$

Then it can be easily shown that s (q) satisfies (2). Indeed, we have

$$\frac{\partial s(q)}{\partial q} b = | Ab \quad b \quad Ab \quad ... \quad A^{n-2} b | = 0,$$

since the first and third columns of this determinant coincide. We next compute

$$\frac{\partial s(q)}{\partial q} A q = \left[\frac{\partial}{\partial q} | A q \ b \ Ab ... A^{n-2} b | \right] A q =$$

$$= [A b \ Ab ... A^{n-2}b] \ Aq = |A^2q \ b \ Ab ... A^{n-2}b|, \tag{9}$$

where [A b Ab ... A^{n-2}b] denotes the vector whole kth component is a determinant of the form | c b Ab ... A^{n-2}b | in which the first column is same as the kth column of matrix A.

It will be now shown that the last determinant in (9) vanishes if q lies on S, that is, if q satisfies Eq. (4). In fact, Eq. (4) implies that there exists a linear combination

$$\lambda Aq + \lambda_0 b + \lambda_1 Ab + ... + \lambda_{n-2} A_{n-2} b = 0 \tag{10}$$

in which at least one of the numbers λ, λ_0, ..., λ_{n-2}, in general, depending on $q \in S$, does not vanish. Pre-multiplication of both sides of Eq. (10) by matrix A yields

$$\lambda A^2 q + \lambda_0 Ab + ... + \lambda_{n-2} A^{n-1} b = 0. \tag{11}$$

Substituting (8) into (11), we find that the vectors $A^2 q$, b, ..., $A^{n-2} b$ are linearly dependent (if, of course, q satisfies (4)). And this implies that the last determinant in (9) vanishes, and consequently $\frac{\partial s}{\partial q} A q = 0$. Thus the assumption that $A^{n-1} b$ is linearly dependent on b, Ab, ..., $A^{n-1} b$ leads to the conclusion that S, given by (4), is an invariant manifold, contradicting the hypothesis of the theorem.

It remains to the shown, as stipulated in Case 2, that the singular manifold S, given by the non-identity (4), is invariant iff the vectors b, Ab,..., $A^{n-2} b$ are linearly independent but the vector $A^{n-1} b$ is linearly dependent on them. But this assertion is simply another way of formulating the theorem just established.

It should be noted that in Case 2, Eq. (4) defining the invariant manifold S can be written in a simpler form

$$\overline{S}(q) \equiv | q \, b \, Ab \, ... \, A^{n-2} b \, | = 0. \tag{12}$$

In fact, since $A^{n-1} b$ is linearly dependent on b, Ab, ..., $A^{n-1} b$, it follows easily from (4) that the subspace S is spanned by these vectors. And it can also be easily verified that the same vectors also span the subspace defined by Eq. (12). Thus in Case 2, Eqs. (4) and (12) are equivalent.

In case 2, thus, the space $\{q\} = R^n$ for system (1) is split into a linear subspace of dimension $(n-1)$, defined by hyperplane (4) or (12), and its orthogonal complement which is a one-dimensional space determined by the normal $\overline{n}$ to this hyperplane. It is known from the general theory [49] that in this case system (1) can be represented as a direct sum of an $(n-1)$-dimensional linear system which is completely controllable in its subspace S, determined by Eq. (4) or (12), and an uncontrollable (in general, independent of u) one-dimensional system in a one-dimensional space determined by the normal vector $\overline{n}$ to the plane (4) or (12). The vector $\overline{n}$ can be expressed in terms of a determinant as follows:

$$\overline{n} = | e \, b \, Ab \, ... \, A^{n-2} b |,$$

where e denotes the column consisting of e_1, ..., e_n, which are the coordinate vectors of the basis of $\{q\}$. In this case, we can say that the space $\{q\}$ has been "foliated". The layers are straight lines parallel to $\overline{n}$. The subspace S, that is, the hyperplane, is a base of the present foliation, and the layers are perpendicular to the base S (Fig. 20.1).

Note that under the given conditions

$$\text{rank} \parallel Aq \quad b \quad Ab \quad ... \quad A^{n-2}b \parallel = n - 1, \tag{13}$$

on account of which (4) and (12) fail to be indentities. Therefore (4) and (12) are equations of the singular manifold of dimension not lower than $n - 1$, and they also guarantee its invariance.

It is also due to condition (13) that in the present case the completion of system (2) continues until the maximum possible number of equations, n, in the system is attained. However, if incontrollability of system (1) were "deeper", that is, if linearly independent were, say, only the vectors b, Ab, ..., $A^{n-3}b$ and the vectors $A^{n-3}b$, $A^{n-1}b$ were linearly dependent on them, the situation would change : there would appear invariant manifold S of lower dimension, namely of dimension $n - 2$, which would be described by two equations $F_1 (q) = 0$, $F_2 (q) = 0$. In this case the extended system (2) already contains $n - 1$ equations of form (3) ; the last equation

$$\frac{\partial F}{\partial q} A^{n-2} b = 0$$

can be dropped since it is a linear combination of the equations

$$\frac{\partial F}{\partial q} b = 0, ..., \frac{\partial F}{\partial q} A^{n-3} b = 0. \tag{14}$$

The condition (4) or (12) is satisfied identically (for any point $q \in \{ q \} = R^n$). This implies that the entire space $\{ q \} = R^n$ constitutes the singular manifold S.

We now attempt to locate directly invariant manifolds of dimension lower than $n-1$. For the case of the linear system (1) it can be easily done. The functions $F_1 (q)$ and $F_2 (q)$ are not uniquely defined although the invariant manifold is, naturally, defined uniquely: this manifold is a subspace spanned by linearly independent vectors

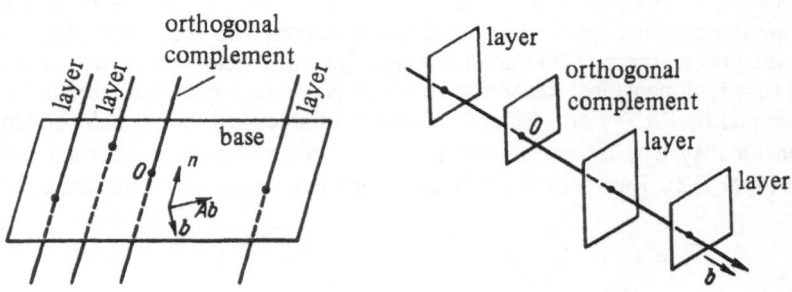

Fig. 20.1   Fig. 20.2

b, Ab, ..., $A^{n-3}b$ It can be treated as "base for foliation". The projection of the representative point on this subspace will be totally controllable in this subspace. The orthogonal complement of this subspace is two-dimensional. It is spanned by two linearly independent vectors $\bar{n}_1$ and $\bar{n}_2$, which can be determined as two linearly independent solutions of the linear homogeneous system

$$\bar{n} b = 0, \quad \bar{n} Ab = 0, \quad ..., \bar{n} A^{n-3} b = 0. \tag{15}$$

Such two solutions of this system exist since the rank of the system of vectors b, Ab, ..., $A^{n-3}b$, and, hence, the rank of the matrix constituted by these vectors is $n - 2$. The layers here are planes parallel to the orthogonal complement. Thus for $n = 3$, for instance, when Ab and A^2b are linearly dependent on b, that is, when both Ab and A^2b are collinear with b, the base is a straight line passing through the origin and collinear with b. The orthogonal complement is a plane perpendicular to b passing through 0. The set of planes orthogonal to b constitutes the set of layers (Fig. 20.2).

Domain of Free Trajectories on Invariant Manifolds

Suppose that there exists an invariant manifold M, with dimension k < n, which the representative point q of a CDS cannot leave under the action of an admissible control. It is interesting to derive conditions under which there exists an open (in the interior metric of M) submanifold $D \subset$ M, also of dimension k, in which any two points $q_0 \in$ D and $q_1 \in$ D can be joined in an admissible way by a trajectory lying entirely in D. We call this submanifold $D \subset$ M the *domain of free trajectories on the invariant manifold* M, in analogy with the notion of the domain of free trajectories which was examined in Sec. 9.

A necessary and sufficient condition for the existence of such a manifold $D \subset$ M is that the cone K (q) coincides with the tangent space T_q M for all $q \in$ M. For the validity of this condition, it is necessary and sufficient, in its turn, that the minimum of the function H (p, q) over all | p | = 1, with p in $\overline{T}_q M$ be strictly positive for all $q \in D \subset$ M. In other words, the condition

$$\varphi(q) = \min (H(p, q) : |p| = 1, p \in \overline{T}_q M) > 0 \qquad \forall q \in D \subset M. \quad (1)$$

must be satisfied. Here $\overline{T}_q M$ is the conjugate space of the tangent space T_q M.

Let us consider an example. Assume that the CDS is governed by the equations

$$\dot{q}^1 = - u^1 q^2, \qquad \dot{q}^2 = u^1 q^1 + u^2 q^3, \qquad \dot{q}^3 = - u^2 q^2,$$

$$|u^1| \le l^1, \qquad |u^2| \le l^2, \qquad l^1 > 0, \qquad l^2 > 0. \quad (2)$$

This CDS has sphere of radius $c \ge O$ as an invariant manifold, since system (2) has a first integral

$$F(q) = \frac{1}{2} [(\dot{q}^1)^2 + (q^2)^2 + (q^3)^2].$$

74

If at a certain instant (initial instant, for example) the representative point q lies on the sphere of a definite radius c, no admissible control $|u^1| \leq l^1$, $|u^2| \leq l^2$, for any $l^1 > 0, l^2 > 0$, can force this point to leave this sphere.

We shall demonstrate that the entire sphere constitutes the manifold D of free trajectories. In fact, we have

$$P(p, q, u) = -p_1 u^1 q^2 + p_2 u^1 q^1 + p_2 u^2 q^3 - p_3 u^2 q^2, \tag{3}$$

and, consequently,

$$H(p, q) = l^1 |-p_1 q^2 + p_2 q^1| + l^2 |p_2 q^3 - p_3 q^2|. \tag{4}$$

The direction orthogonal to $T_q D$ is determined by the unique vector normal to the sphere at q :

$$\frac{\partial F}{\partial q} = (q^1, q^2, q^3)^T = q. \tag{5}$$

Relation (4) shows that $H(p, q)$ is, in any case, non negative. It can vanish only if $|-p_1 q^2 + p_2 q^1| = 0$ and $|-p_2 q^3 + p_3 q^2|$ simultaneously. This takes place, as can be easily verified, only if $p \neq 0$ is collinear with q. But p q = 0, in view of (1). Consequently, p and q are noncollinear. Thus $\varphi(q) > 0$ for any point q of the sphere, which is what we set out to establish.

Thus CDS (2) has as an invariant manifold a sphere of arbitrary (but fixed) radius which at the same time constitutes a manifold of unconstrained (free) trajectories.

To conclude the present section, it is pertinent to remark that, since the lateral boundary of trajectory funnel does not always exist, we must investigate in this case the nature of admissible trajectories passing through a given point, q'. Generally speaking, this is quite a difficult problem if one has to consider cones $K(q)$ of arbitrary type a_r^π in the neighbourhood of q'. Hence we shall examine only one, not the most complex, case.

Consider the CDS governed by the equation

$$\dot{q} = u^1 f_1(q) + u^2 f_2(q),$$
$$q \in D \subset R^3, \quad |u^{1,2}| \leq 1, f_i \in C^\infty, \tag{6}$$

where D is a given domain and q' is an interior point of D. In D, all the cones $K(q)$ of the field of cones are of the same type a_0^2, that is, they are disks. By means of a piecewise constant control u(t) it is possible to perform successive motion along the vectors f_1 and f_2. In particular, if u(t) is defined over the time interval $[0, 4\sqrt{t}]$ by the formulas

$$\begin{cases} u^1(\tau) = 1 \\ u^2(\tau) = 0, \tau \in [0, \sqrt{t}], \end{cases} \qquad \begin{cases} u^1(\tau) = 0 \\ u^2(\tau) = 1, \tau \in [\sqrt{t}, 2\sqrt{t}], \end{cases} \tag{7}$$

$$\begin{cases} u^1(\tau) = -1 \\ u^2(\tau) = 0, \tau \in [2\sqrt{t}, 3\sqrt{t}], \end{cases} \qquad \begin{cases} u^1(\tau) = 0 \\ u^2(\tau) = -1, \quad \tau \in [3\sqrt{t}, 4\sqrt{t}], \end{cases}$$

then in time $4\sqrt{t}$ the system moves under the action of the given control from the point q' to the point q (t) given by

$$q(t) = q' + t[f_1, f_2] + O(t^{3/2}), \qquad (8)$$

here $O(t^{3/2})$, denotes an infitesitimal of order $t^{3/2}$ [90a].

We now consider, over the time interval $[0, t_1 + t_2 + 4\sqrt{t_3}]$, the motion of (6) from the given point q' under the action of the control u (t) defined as follows :

$$\begin{cases} u^1(\tau) = 1 \\ u^2(\tau) = 0, \quad \tau \in [0, t_1], \end{cases} \qquad \begin{cases} u^1(\tau) = 0 \\ u^2(\tau) = -1, \quad \tau \in [t_1, t_1 + t_2], \end{cases} \qquad (9)$$

and for $\tau \in [t_1 + t_2, t_1 + t_2 + 4\sqrt{t_3}]$ the control u (t) is defined by (7) with $t = t_3$. This motion results in q' moving over to the point q (t_1, t_2, t_3) which depends on the values t_1, t_2, t_3 of In this way, we obtain a mapping q (t_1, t_2, t_3) of a set of values of the parameters t_1, t_2, t_3 into the state space of the CDS.

Denoting, respectively, by $\varphi(\tau, q')$ and $\psi(\tau, q')$ solution of the equations with the initial point q', for the transformation q (t_1, t_2, t_3) we obtain the expression

$$q(t_1, t_2, t_3) = t_3[f_1, f_2] + O(t_3^{3/2} + \psi(t_2, \varphi(t_1, q'))). \qquad (10)$$

The derivatives, at $t = 0$, of $\varphi(\tau, q')$ and $\psi(\tau, q')$ with respect to the initial data are of the form

$$\frac{\partial \varphi^i}{\partial q^j}(0, q') = \delta_j^i, \qquad \frac{\partial \psi^i}{\partial q^j}(0, q') = \delta_j^i, \qquad i, j = 1, 2, 3, \qquad (A)$$

since $\varphi(0, q') = q'$, $\psi(0, q') = q'$. Therefore, nothing q (0) = q', for the Jacobian matrix of the transformation q (t_1, t_2, t_3) we have at the point (t_1, t_2, t_3) = 0

$$\frac{\partial q}{\partial \theta} = (f_1(q'), f_2(q'), [f_1, f_2](q')) \qquad (B)$$

for $\theta = (t_1, t_2, t_3)$ and $\theta = 0$. In view of the theorem on continuous differentiability of solutions of differential equations with respect to initial data, the Jacobian of the transformation is a continuous function of $\theta = (t_1, t_2, t_3)$. Thus, provided

$$|f_1(q'), f_2(q^1), [f_1, f_2](q')| \neq 0, \qquad (11)$$

the transformation q (t_1, t_2, t_3) has the maximum rank in a neighbourhood of the point θ - 0. According to the implicit function theorem [30 a], there exiests a domain D_1 containing the point $\theta = 0$ such that the transformation q (θ) is a diffeomorphism of D_1 onto an open region D_2 containing q'. This enables us to state the following.

If the condition (11) is satisfied at the point q', then an $\in$ > 0 can be found such that for any $\delta < \in$ there exists an h > 0 with the property that any q, which satisfies l q - q' l < δ, is accessible from q' in time

$$T \leq 2h\ (\ \delta\ ) + 4\ h\ (\ \delta\ ) = 6h\ (\ \delta\ )\ \text{where}\ \lim_{\delta \to 0} h\ (\ \delta\ ) = 0.$$

Here h (δ) can be treated as the maximum time of motion along trajectories of the fields f_1 (q) and f_2 (q) between the points of intersection of these curves with the boundary l q - q' l = δ of the domain l q - q' l < δ. That is, h (δ) is the maximum time of possible motion along at least one of the integral curves of the field f_1 (q) or f_2 (q) during which it is not possible to leave the boundary of the domain l q - q' l < δ. The condition $\lim_{\delta \to 0} h\ (\ \delta\ ) = 0$ is a

consequence of the fact that the transformation q (t_1, t_2, t_3) is a diffeomorphism. Thus, in the present case, the point q' cannot be the vertex of the rigid trajectory funnel (it does not exist). And depending on whether or not condition (11) is satisfied, we have either a domain of local controllability near q' or else through q' there passes a two-dimensional invariant manifold. In case of the invariant manifold, near this point there will be a domain of unconstrained motions on the manifold.

22

Trajectories Funnel on Invariant Manifolds

Let the CDS

$$\dot{q} = f(q, u), \quad u \in U, \quad q \in \{q\} = R^n, \quad (1)$$

have an m-dimensional smooth invariant manifold M given by the equations

$$F^k(q) = 0, \quad k = 1, 2, ..., n - m. \quad (2)$$

The mainfold M lies in $\{q\}$ in a natural way or, alternatively, as is customary to say, is embedded in $\{q\}$. At every point $q \in M$ there is a plane tangent to M which, as mentioned earlier, is called the *tangent space* to M at q, and is denoted by T_qM. In other words, we can say that T_qM is the orthogonal complement of the linear space spanned by the vectors $\partial F^k/\partial q$, $k = 1, 2, ..., n - m$. Thus, since M is invariant with respect to the CDS (1),

$$\frac{\partial F^k}{\partial q} \dot{q} = 0, \quad k = 1, 2, ..., n - m, \quad (3)$$

for arbitrary controls $u \in U$ and $q \in M$ [1]. Consequently all the admissible vectors $\dot{q}$ lie in T_qM. That is, $\dot{q}$ are vectors tangent to M at q. It should be remarked that the tangent vectors to M at q can be defined directly as well in terms of "internal" properties of M without invoking the fact that M is embedded in $\{q\}$.

We shall assume that the point $q' \in M$ and its neighbourhood in M do not lie in the submanifold of free trajectories on M (Sec. 21) and that they lie in the relative interior of M. Then the cone $K(q')$ of admissible velocity directions of CDS (1), which lies completely in $T_{q'}M$, does not coincide with $T_{q'}M$. We assume further that in M there are no other invariant manifolds of the CDS of dimension lower than m. Under these conditions, the trajectory funnel $V(q')$, which lies wholly in M, will have dimension m.

---

[1] If M has an edge, it is assumed that q does not lie on the edge.

78

The trajectory funnel V (q')[1] is, at least in the neighbourhood of q', a conoid of dimension m lying wholly in the invariant manifold M. As in Sec. 14, the lateral boundary ∂V (q') of this conoid, if it exists, is given by the parametric equations

$$q = q (t, q', p'), \qquad q (0, q', p') = q', \qquad (4)$$

where the parameters are $t > 0$ and p'; $q' \in M$ being fixed.

The function q (t, q', p') is obtained as a solution of the Hamilton's system of equations

$$\dot{q} = \frac{\partial H}{\partial p}, \qquad \dot{p} = -\frac{\partial H}{\partial q}, \qquad (5)$$

and p' runs through not the entire cone $\overline{K}$ (q'), the conjugate of K (q') (as in Sec. 14), but only through its subset defined by the condition

$$p' \left(\frac{\partial F^k}{\partial q} \right)^T = 0, \qquad k = 1, 2,..., n - m. \qquad (6)$$

Condition (6) can also be interpreted in a different manner: p' must run through the cone $\overline{K}_1$ (q'), which is the projection of $\overline{K}$ (q'), onto the conjugate space of $T_q \cdot M$. This space is called the *cotangent space* of M at $q' \in M$, and is denoted by $\overline{T}_d M$.

Just as the case where CDS (1) did not have any invariant manifold, we can determine hatching of the exterior side of the boundary of the trajectory funnel V (q') $\subset$ M as well as singular submanifolds S $\subset$ M (and consequently submanifolds of reverse hatching) and, possibly, new invariant submanifolds $M_1 \subset M_0$.

We conclude the present section with a detailed analysis as to in what way does the Hamilton's system (5) admit of contraction on the invariant manifold M. In this case, for all $q \in M$ the domain of admissible values of the covector p in the contangent space $\overline{T}_q M$ of M at q. The contracted Hamiltonian is of the form

$$H_1 (p, q) = \underset{u \in U}{\text{Sup (max)}}\, p\, f (q, u), \quad q \in M, p \in \overline{T}_q M. \qquad (7)$$

This follows from (3). Indeed, substituting $\dot{q} = \partial H / \partial p$ into (3), we have

$$\frac{\partial F^k}{\partial q} \cdot \frac{\partial H}{\partial p} = 0, \qquad k = 1, ..., n - m ; \qquad \forall p ; \qquad \forall q \in M. \qquad (8)$$

This implies that for $q \in M$, the Hamiltonian H (p, q) does not depend on the components of the covector p which lies in the linear space spanned by the covectors $\partial F^k / \partial q, k = 1, ..., n - m$. That is.

---

[1] The trajectory funnel is assumed to be rigid (Sec.14).

$$H(p+p_1,q) = H(p,q), \quad \forall q \in M; \; \forall p_1 \in \overline{T}_q M. \tag{9}$$

Thus in order to construct the trajectory funnel $V(q')$, $q' \in M$, it is important to know, at the point $q(t) \in M$, only the projection of the impulse $p(t)$ into the cotangent space $\overline{T}_{q(t)} M$. This justifies the transition to the Hamilton's system on M with the Hamiltonian defined by (7).

Let us now construct the contracted Hamilton's system on M in terms of local coordinates on M. Suppose that in the neighbourhood of any given point $q \in M$ there are defined coordinates x_j, $j = 1, ..., m$, on M. Suppose further that $\partial F^k / \partial q \neq 0$, $k = 1,..., n - m$, and that the vectors $\partial F^k / \partial q$ are linearly independent. Then in the neighbourhood of q there exists a nondegenerate coordinate transformation

$$\begin{aligned} x^j &= x^j(q), & j &= 1, 2, ..., m, \\ F^k &= F^k(q), & k &= 1, 2, ..., n - m. \end{aligned} \tag{10}$$

Under the transformation, the vectors lying in the tangent space of R^n are transformed according to the formulas

$$\overset{.}{x}{}^j = \frac{\partial x^j}{\partial q} \overset{.}{q}, \qquad \overset{.}{F}{}^k = \frac{\partial F^k}{\partial q} \overset{.}{q}, \tag{11}$$

while the covectors according to the formulas

$$p_i = \pi_j \frac{\partial x^j}{\partial q_i} + \theta_k \frac{\partial F^k}{\partial q_i} \tag{12}$$

where π_j, θ_k are covectors belonging to subspaces which are conjugate of subspaces containing $\overset{.}{x}{}^j$, $\overset{.}{F}{}^k$. In the new coordinates, the equations of the original CDS (1) become

$$\overset{.}{x} = \zeta(x, F, u), \qquad \overset{.}{F} = \eta(x, F, u), \tag{13}$$

where, in accordance with (11),

$$\zeta(x, F, u) \equiv \frac{\partial x}{\partial q} f(q, u), \tag{14}$$

$$\eta(x, F, u) \equiv \frac{\partial F}{\partial q} f(q, u), \tag{15}$$

The invariance condition of M implies that

$$\eta(x, o, u) = 0. \tag{16}$$

On account of (11) and (12), the Hamiltonian H (p, q) in (x, F) - coordinates becomes

$$H (p, q) = \sup_{u \in U} (\max) \, pf (q, u) =$$

$$= \sup_{u \in U} (\max) \left[\left(\pi_j \frac{\partial x^j}{\partial q} + \theta_k \frac{\partial F^k}{\partial q} \right) (q, u) \right] =$$

$$= \sup_{u \in U} (\max) [\, \pi_j \zeta^j (x, F, u) + \theta_k \eta_k (x, F, u)] = H_2 (\pi, \theta, x, F), \quad (17)$$

where the summation is performed over repeated indices. The Hamilton's equations are of the form

$$\dot{x}^j = \frac{\partial H_2}{\partial \pi_j}, \qquad \dot{\pi}_j = - \frac{\partial H_2}{\partial x^j} \qquad j = 1, ..., m,$$

$$\dot{F}^k = \frac{\partial H_2}{\partial \theta_k}, \qquad \dot{\theta}_k = - \frac{\partial H_2}{\partial x^k} \qquad k = 1, ..., n - m, \qquad (18)$$

The solution of this system, lying in the coordinate space on M, satisfies the condition $F (t) \equiv 0$. To see this, consider the solution with an arbitrary given initial point on M : $(p^0, q_0) = (\pi^0, \theta^0, x_0, 0)$, that is, $q_0 \in M$, $p^0 \in \bar{T}_{q_0} R^n$. Nothing (16), (17), we find that for $F = 0$ the Hamiltonian $H_2 (\pi, \theta, x, F)$ is of the form

$$H_2 (\pi, \theta, x, 0) = \sup_{u \in U} (\max) [\pi_j, \xi^j (x, 0, u)] = H_3 (\pi, x). \qquad (19)$$

Hence

$$\left. \frac{\partial H_2}{\partial \theta_k} \right|_{F = 0} = \frac{\partial}{\partial \theta_k} \sup_{u \in U} (\max) [\, \pi_j \xi^j (x, 0, u)] \equiv 0.$$

implying $F (t) \equiv 0$, in view of (18). The solution of the Hamilton's system (18) with the initial point (p^0, q^0) (that is, $q_0 \in M$, $p^0 \in \bar{T}_{q_0} R^n$) thus satisfies the following system :

$$\dot{\theta} = - \left. \frac{\partial H_2}{\partial F} \right|_{F = 0} \qquad (20)$$

$$\dot{F} = 0, \qquad (21)$$

$$\dot{\pi} = -\frac{\partial H_3}{\partial x} \tag{22}$$

$$\dot{x} = \frac{\partial H_3}{\partial \pi} \tag{23}$$

Eqs. (22) - (23) constitute a Hamilton's system, with the Hamiltonian (19), for the CDS

$$\dot{x} = \xi(x, o, u), \quad u \in U, \quad x \in R^\pi \tag{24}$$

In other words, Eqs. (22), (23) constitute a Hamilton's system contracted on M, while the Hamiltonian $H_3(\pi, x)$ is the contraction on M of the Hamiltonian $H(p, q)$ expressed in terms of the local coordinates on M.

Let $\mu : R^n \longrightarrow M$ denote the projection of R^n onto M. This projection, in (x, F) – coordinates, is of the form $\mu(x, F) = x$. Likewise, we denote $\mu_q^* : \overline{T}_q R^n \longrightarrow \overline{T}_{\mu(q)} M$. In view of (20), (21), (22) and (23), we have established the following

Theorem. For any given point $q_0 \in M$, $p^0 \in \overline{T}_{q_0} R^n$ and the corresponding point $\tilde{q}_0 = q_0 \in M$, $\tilde{p}^0 = \mu_{q_0}^* p^0$, the solution $\tilde{q}(t), \tilde{p}(t)$ of the Hamilton's system contracted on M is the projection of the solution of the original Hamilton's system locally on M. That is,

$$\exists \varepsilon(q_0) > 0 : \forall q : |q - q_0| < \varepsilon(q_0), \tilde{q}(t) = q(t, \tilde{p}(t)) = \mu_{q(t)}^* p(t).$$

It should he noted that, in general, $\dot{\theta}(t) \neq 0$, in view of (20). This implies that if the impulse covector $p^0 \in \overline{T}_{q_0} R^n$ at the initial moment, then during subsequent moments the covector $p(t)$ will not, in general, belong to $\overline{T}_{q(t)} M$.

Separating Hypersurfaces in State Spaces

Suppose that (Sec. 6) the CDS governed by the connection

$$\dot{q} \in f(q, U), \tag{1}$$

is uniquely characterized by the continuous field of cones K (q). In a domain $G \subset \{ q \}$ there can exist two-sided hypersurfaces R, called *separating* hypersurfaces, with the following property : for $q \in R$, the cone K(q) lies entirely on one side of R. We shall refer to this side as the *internal side* of R. Let us perform hatching on the external side of R. Hatched separating hypersurfaces, if they exist, carry information regarding admissible trajectories of CDS in $G \subset \{ q \}$ which can prove very useful in investigating the given CDS.

It follows from the definition of R that if an admissible trajectory crosses R transversally, it can do so only from the side of hatching, and not from the opposite side. Such a hatched surface is analogous to the hatched surface of a rigid trajectory funnel, itself a limiting case of R.

It is easy to derive a necessary condition to be satisfied by R. Let z(q) = 0 be the equation of a smooth separating surface R in G, and let $\partial z / \partial q$ be the outward normal to R. Then obviously the inequality

$$H\left(\frac{\partial z}{\partial q}, q\right) < 0 \qquad \forall\, q \in R, \tag{2}$$

where H (p, q) is the Hamiltonian of CDS (1), is satisfied. Inequality (2) can be turned into an equality. In fact, (2) shows that whenever a surface R exists, so does a positive function ψ (q)

$$H\left(\frac{\partial z}{\partial q}, q\right) + \psi(q) = 0. \tag{3}$$

This is the desired equation to be satisfied by z(q), which determines the equation of the required R. We observe that the inequality (2) is equivalent to the inclusion

$$\frac{\partial z}{\partial q} \in \mathrm{ri}\, \overline{K}\, (q),$$

(4)

where $\overline{K}\, (q)$ is the conjugate cone of $K\,(q)$ (Sec.4).

In general, Eq. (3) is the usual nonlinear partial differential equation of first order in unknown z(q). It can be solved by standard techniques ; in particular, by the method of characteristics. Characteristics are given by canonical equations with the Hamilton function

$$H'(p, q) = H(p, q) + \psi(q).$$

(5)

In the present case, these equations are of the form

$$\dot{q} = \frac{\partial H'}{\partial p} = \frac{\partial H}{\partial p}\,,$$

(6)

$$\dot{p} = -\frac{\partial H'}{\partial q} = -\frac{\partial H}{\partial q} - \frac{\partial \psi}{\partial q}\,.$$

Eqs. (6) differ from characteristic equations describing the surface of integral funnel only in the second equation (for impulses) where a new term $\frac{\partial \psi}{\partial q} = \mathrm{grad}\, \psi\,(q)$ appears. The presence in (5) of the additional term, in essence at our disposal ($\psi\,(q)$ is a positive function), can greatly simplify the task of solving Eq. (3) or (6).

The solution of Eq. (3) depends on a positive function $\psi\,(q)$, and from the set of these solutions a family of hatched surfaces R can be chosen which contains information regarding admissible trajectories in the whole space { q } or in a domain G of it (Fig.23.1). For instance, if at least one R has been found which, firstly, is hatched entirely from one side and, secondly, divides { q } into two disjoint parts, then it can be concluded that the given CDS is uncontrollable.

To consider an example, let the given CDS be of the form

$$\dot{q} = a(q) + b(q, u),\qquad\qquad u \in U(q),$$

(7)

with

$$|a(q)| > \max |b(q, u)| + \varepsilon,$$

$$\forall\, q \in \{q\}\,, \varepsilon > 0.$$

(8)

The condition (8) implies that the cone $K\,(q)$ does not coincide with $\{\dot{q}\}$ for any $q \in \{q\}$. It is assumed further that a (q) is a potential vector, that is, there exists a scalar function w (q) such that

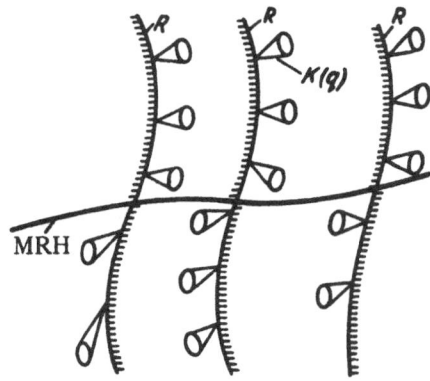

Fig. 23.1

$$a(q) = \frac{\partial w}{\partial q}(q) = \text{grad } w(q).\tag{9}$$

It follows from condition (8) that when q lies on the surface $w(q) = c$, the cone $K(q)$ lies on one side of this surface. If we now assume that the surface $w(q) = c$ divides the space into two disjoint parts (not at all a rare case in practice), then (7) becomes an uncontrollable system. Thus (8) and (9) are sufficient conditions for an arbitrary CDS to be uncontrollable.

Apart from separating surfaces R, considered in the present section, it is useful also to examine a surface which can be called *trajectory raving* (or *trajectory wedge*) of the differential inclusion (1) and the associated CDS. To explain this notion, we consider the case n = 3 ; generalization to the case n > 3 is obvious.

A trajectory ravine for CDS (1) in the case n = 3 is introduced as follows.

We take in { q } a non-selfintersecting curve 1 with the following property. If $q \in 1$, then in the neighbourhood of q the curve 1 does not go inside and does not touch the cone $K(q)$. With every $q \in 1$ as vertex, we draw a characteristic conoid (the surface of the rigid trajectory funnel $V(q)$ with vertex q). This results in a family of conoids $V(q)$ with parameter $q \in 1$. We now draw enveloping surfaces for this family. The surface thus obtained is called the *trajectory ravine (wedge)* of the given CDS. The curve 1 of tapering serves as the base of the ravine or the edge (rim) of the wedge.

It is plain that a trajectory ravine is a limiting case of the separating surface R. The equation of a trajectory ravine $z(q) = 0 (z(1) = 0)$ is such that $z(q)$ satisfies Eq. (3) with $\psi(q) = 0$. That is, $z(q)$ satisfies the same equation which is satisfied by the boundary of a trajectory funnel.

24

Admissible Manifolds of CDS in State Space

Another kind of manifolds can be considered in the state space. These too prove useful in the investigation of a given CDS.

An m-dimensional manifold M^m will be referred to as an *admissible manifold* if there exists an admissible control $u(t)$, $t_0 \leq t \leq t + \epsilon$, $\epsilon > 0$, such that the representative point $q(t)$, $q(t_0) \in M^m$, belongs to M^m, that is, $q(t) \in M^m$ for $t_0 \leq t \leq t + \epsilon$, $\epsilon > 0$. If such an admissible manifold M^m exists, then we can consider the motion of the given CDS only on M^m, and in this sense speak of *contraction* or *restriction of the given CDS to the manifold* M^m.

The given CDS restricted to M^m can be treated as a new CDS with its own Hamiltonian $H'(p, q) = \sup(\max) p \dot{q}$, where for each $q \in M^m$ the upper bound (maximum) is taken over all $\dot{q}$ lying in the intersection of $T_q(M)$ and $\Phi(q)$ of the original CDS goverened by the inclusion $\dot{q} \in \Phi(q)$.

In other words, an admissible manifold is a manifold along which motion can be performed by means of admissible controls. In a sense, the notion of an admissible manifold supplements that of a separating manifold.

It is clear that an invariant manifold of the given CDS is simultaneously an admissible manifold. But the converse as not always true. Also clear is the fact that an admissible trajectory is an admissible manifold of dimension $m = 1$.

A necessary and sufficient condition for a manifold M^m to be an admissible manifold is that the tangent hyperplane has $T_q M^m$ has a non-empty intersection (other than the point $\dot{O}(q)$ with the cone $K(q)$ of admissible velocities of CDS for all $q \in M^m$ 13).

Let M^m be given by the equations

$$F^k(q) = 0, \qquad k = 1, ..., \quad n - m. \tag{1}$$

Then for all $q \in M$ there exists a $u \in U$ such that

$$\frac{\partial F^k}{\partial q}(q)f(q,u)=0, \qquad\qquad k=1,\ ...,\ n-m. \qquad\qquad (2)$$

Condition (2) also provides a sufficient condition for the manifold defined by Eqs. (1) to be an admissible manifold. It follows from (2) that

$$\sup_{u\in U}\ (\text{max})\frac{\partial F^k}{\partial q}\ f(q,u)=H\left(\frac{\partial F^K}{\partial q},q\right)\ \geq\ 0 \qquad\qquad (3)$$

as well as

$$\inf_{u\in U}\ (\text{min})\frac{\partial F^k}{\partial q}\ f(q,u)=H\left(\frac{\partial F^k}{\partial q},q\right)\ \geq\ 0 \qquad\qquad (4)$$

Conditions (3), (4), in turn, imply that for the existence of an admissible manifold it is necessary that there the nonnegative functions ψ_1 (q) and ψ_2 (q) ensuring the existence of a nontrivial (nonzero) solution of the system of two partial differential equations (nonhomogeneous, in general)

$$H\left(\frac{\partial F}{\partial q},q\right)-\psi_1(q)=0, \qquad H\left(-\frac{\partial F}{\partial q},q\right)-\psi_2(q)=0, \qquad\qquad (5)$$

in the unknown function F (q).

25

Phase Portrait of CDS

At the present time, it is apparently difficult to give a general and exhaustive definition of the phase portait of CDSs considered here. At the present stage of development of the theory, it can be adequately done for a second-order CDS defined on the plane or on a two-dimensional manifold (Sec. 34). In the general case, by the phase portrait of a CDS we mean an aggregate of geomatric and analytic aids for representing the CDS that give as far as possible a complete, clear and vivid description of admissible trajectories, domains of reachability and controllability as well as other characteristic features of the given CDS. The aids and notions of a geometrical nature will be referred to as *elements* of the phase portrait of the CDS. It will be useful to include among these aids a number of notions introduced in the present book such as the Hamiltonian and the Lagrangian of the CDS, types of cones of CDS, trajectory and integral funnels and their hatched surfaces, separating surfaces, admissible surfaces, singular surfaces of reverse hatching, invariant surfaces etc. These notions serve as examples of elements of the phase portrait of a CDS.

Whether or not all elements of the phase portrait are made use of depends on a particular problem for the given CDS ; it may happen that information furnished is more than what we actually need.

One of the devices of constructing the phase portrait of a CDS consists in decomposing the state space of CDS into disjoint sets of fixed type. Every such set is characterized by the fixed type a_r^m of the cone K (q). Analytically or geometrically form and boundaries of these sets are determined. Depending on the nature of integrability of Pfaffian differential forms, defined by Eqs. (8.8), (8.9), a set of type a_r^m is decomposed into subsets. Here it proves to be useful to switch over to the dual description in terms of partial differential equations. This way we can ascertain the presence of, say, invariant manifolds, conoidal surfaces of rigid trajectories of funnels etc. It is clear that in the general case, the question of ascertaining the nature of admissible trajectories, reachability domains etc. may turn out to be nontrivial even in the case of fixed type a_r^m . On the other hand, however, it is also true that, for example, for domains where the cone is of type a_r^m , that is, for domains of free (unconstrained) trajectories (Sec.9), the problem of ascertaining the nature of admissible trajectories is very simple.

The study of the nature of admissible trajectories and motions of CDS on sets of fixed type a_r^m and its subsets can be termed as the local investigation of CDS. Such an investigation corresponds to "differentiation" of the phase space of CDS on subsets with sufficiently "homogeneous" structure.

The local investigation in the nature of the phase portrait of CDS is naturally followed by the question of global investigation. That is, the question of a sort of "integration" of separate elements of the phase portrait into one complete picture. This picture can be represented in the form of a graph. Here the problem which arises concerns the existence and description of admissible trajectories that join points lying in adjacent sets of fixed type and their subsets.

Possible transitions of the representative point along admissible trajectories between various sets and subsets may be represented in the form of a graph whose vertices depict sets and subsets of points of the state space and edges depict possible transitions. Such a graph enables us to view controllabillity and reachaibility properties of CDS in its entire state space, that is, view the phase portrait of CDS in its entirety (in the large).

The method of constructing phase portrait of CDS suggested here is realized in large measure for two-dimensional CDSs with state space in the form of a two-dimensional plane or a two-dimensional manifold (Secs. 34-40).

26

Transient and Relative Motion of CDS

A CDS is characterized by the convex set f (q, U) in the velocity space { $\dot{q}$ }. Choose a point O´(q) in the relative interior ri f (q, U). Then the set f (q , U) can be obviously expressed as the sum (Fig. 26.1)

$$f (q , U) = a (q) + b (q , U) , \tag{1}$$

where a (q) is a vector independent of u ∈ U and b (q , U) is a convex set containing the origin $\dot{O}$ (q) of the space { $\dot{q}$ }. The set b (q , U) is obtained as the set of vectors b (q , U) ≡ f (q , U) – a (q) when u ranges over all of U. We shall call a (q) the *velocity vector of transient* (uncontrollable) *motion*. We call b (q , U) the *set of admissible velocities of relative motion*. The term "relative" here is conditional since b (q , U) depends on the point q of the "absolute" space { q }.

Such a representation of an arbitrary given set f (q , U) can prove to be very useful in solving concrete problems [117, 118]. In the optical analogy of CDS (Sec. 30) such a representation can be interpreted as a source moving with velocity a (q) which, in turn, in relative motion radiates all sides.

Assume, for example, that b (q , U) is an n-dimensional sphere of radius c (q) > 0 with centre $\dot{O}$ (q). If c is interpreted as the maximum velocity of perturbation propagation in the given medium (velocity of light or sound), then for | a (q) | < c (q) we have motion of the source with velocity less than that of light (subsonic motion) and the wave fronts will be of the form of embedded spheres (Fig. 26.1a). For | a (q) | > c (q) we have motion of the source with velocity greater than that of light (supersonic motion) and the envelope of wave fronts will be a conoid (surface of a

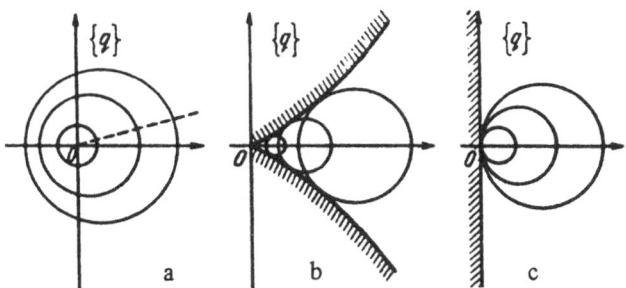

Fig. 26.1

trajectory funnel) (Fig. 26.1b). The characteristic conoid depicted in Fig. 26.1b is nothing but the *light conoid* (in particular, *light cone*) of the theory of relativity considered in the event space { q , t } [17 , 115]. Finally, we have the border case I a (q) I = c (q) (Fig. 26.1c). In the present example, the domain of unconstrained trajectories of the corresponding CDS is determined by the condition I a (q) I < c (q), while the points of the boundary of this domain satisfy the equation I a (q) I = c (q).

It is interesting to examine a not so uncommon particular case of a division of motions where b (q , u) or b (q , U) is independent of q. That is, admissible velocity of a controllable motion does not depend on the position of the moving source in { q }. In this case the CDS is of the form

$$\dot{q} = a(q) + u, \tag{2}$$

where u ∈ U and U is a convex set containing $\dot{O}$ (q) ∈ { $\dot{q}$ } and is independent of q. For CDS (2), the corroresponding cone K (q) of admissible directions does not change its form but changes only with q. In this case H (p , q) is of the form

$$H(p,q) = \max_{u \in U} (\sup) [pa(q) + pu] = pa(q) + h(p), \tag{3}$$

where h (p) is the support function of U and is independent of q. The canonical (characteristic) equations assume the form

$$\dot{q} = \frac{\partial H}{\partial p} = a(q) + \frac{\partial h}{\partial p}(p).$$

$$\dot{p} = -\frac{\partial H}{\partial p} = -p\frac{\partial a}{\partial q}(q). \tag{4}$$

A sailent feature of Eqs. (4) is that the right side of the first equation is a sum of two functions each of which depends only on q or only on p and the second equation is linear in p.

27

CDS with Ellipsoidal Indicatrix

Consider the CDS

$$\dot{q} \in f(q, U),\tag{1}$$

where the convex set of admissible velocities $f(q, U)$ of dimension n in the space $\{\dot{q}\} = R^n$ is an ellipsoidal convex set. In other words, the indicatrix set of CDS (1) is described by the inequality

$$\frac{1}{2}\dot{q}^T A \dot{q} + (a, \dot{q}) + \alpha \le 0,\tag{2}$$

where A is a given n x n positive definite symmetric matrix depending, in general, on q, and a and α are given vector and scalar, respectively, possibly also depending on q. It is easy to determine the support function or the Hamiltonian associated with (2) ; it is

$$H(p, q) = (2 \beta p A^{-1} p^T)^{\frac{1}{2}} + pb,\tag{3}$$

where

$$b = -A^{-1} a^*, \beta = \frac{1}{2} a^T A^{-1} a^* - \alpha.\tag{4}$$

These formulas together with (9.2) enable us to determine the domain of motions. It is given by the inequality

$$\varphi(q) = \min_{|p|=1} H(p, q) = \min_{|p|=1} [(2 \beta p A^{-1} p^T)^{\frac{1}{2}} + pb] > 0.\tag{5}$$

92

On the other hand, it follows directly from (2) that this domain is determined by the condition

$$\alpha = \alpha(q) > 0. \tag{6}$$

CDSs with domain of admissible velocities of form (2) are interesting in their own right, as also from the point of view that an arbitrary bounded convex set may be approximated by a set of form (2) to a given degree of accuracy. What is interesting is that the dimension of the set which is being approximated can be less than n. For example, if a set of a CDS of dimension 2 constitutes a segment (that is, the scalar control u^1 is bounded in magnitude : $|u^1| \leq 2l$), then it can be approximated by an ellipse

$$\frac{(u^1)^2}{4l^2} + \frac{(u^2)^2}{b^2} = 1 \tag{7}$$

on the plane (u^1, u^2). When $b \longrightarrow 0$, the ellipse (7) will "tend" to the segment $|u^1| \leq 2l$.

Such approximations can prove to be very useful [117,118]. For example, if, subject to (7), the controllability of the CDS is established and is shown that for $b \longrightarrow 0$ there exist admissible controls realizing this controllability by means of a uniform limiting process bounded above by time, then the "limiting" CDS, which coincides with the original CDS, will also be, in general, controllable.

CDS and Continuous Non Linear Media. The Principle of Maximum of Flows of Substance. Laplace Operator of CDS

In the present section we shall show that with every CDS governed by the inclusion $\dot{q} \in f(q, U)$, $q \in \{q\}$ a definite continous medium can be associated in which there takes place a propagation to a substationof, for example, heat, mass, charge etc. To be definite, we shall consider a heat transmitting stationary medium which occupies an n-dimensional space $\{q\} = R^n$. The state of this medium will be described by a function of temperature distribution $z = z(q, t)$, where $q \in \{q\}$ and t denotes the time. For brevity, in the sequel we shall drop the argument t in writing this function.

It is known [52] that properties of such a medium are described by an operator A (grad z (q) , q). This operator associates the gradient of temperature $\partial z / \partial q = \text{grad } z(q)$ with the flow vector K (q) :

$$K(q) = A(\text{grad } z(q), q), q \in \{q\}. \tag{1a}$$

In the field theory, relations such as (1a) are known as *material relations*. In the classical theory [52], A is usually an algebraic linear operator which associates K (q) with grad z (q).

However, there are continuous media where material relations are not linear. Such media are called *inhomogeneous*. We come across *inhomogeneous* media in several natural and technical objects. To cite few examples, these are layered media, active media with negative propagation and energy conservation, and many other forms of continuous media [30, 64, 72, 99].

The aim of the present section is, in particular, to make precise relations (1a). This consists in relating the quantities K and grad z by means of *"the principle of maximum flow"*. Before we are able to formulate this principle, we have to introduce some necessary concepts.

First of all, we introduce the set $\Phi(q)$ of points of the tangent space $T_q \{q\}$.

This set is defined for each fixed $q \in \{q\}$. To make our discussion simplified, we assume that $\Phi(q)$ is a non-empty, bounded, closed and strictly convex set. $\Phi(q)$ will be called the *set of admissible directions of flow*.

94

The condition that Φ (q) is strictly convex has been imposed in order to ensure that K (q) is uniquely defined from the principle of maximum flow formulated below. It can be shown that for our purpose this condition is not of principal importance and can be replaced by the condition that Φ (q) is simply convex. However, we shall not discuss this question here. The flow vector K (q) is assumed to lie in T_q { q } and to originate from the point $\dot{O}$ (q). If the endpoint of K (q) lies in Φ (q), we shall write K (q) $\in$ Φ (q) or K $\in$ Φ (q).

THE PRINCIPLE OF MAXIMUM FLOW. *Assume that the vector* K_m *(q) satisfies the following maximum condition*

$$\max_{\kappa \,\in\, \Phi(q)} \; [\text{grad } z\,(q)\,.\,\kappa\,] \;=\; \text{grad } z\,(q)\,.\,\kappa_m\,(q), \qquad a \,\forall\, q \in \{\,q\,\}, \tag{1}$$

or, what is same,

$$\kappa_m\,(q) \;=\; \arg\max_{x \,\in\, \Phi(q)} \; [\text{grad } z\,(q)\,.\,\kappa\,], \qquad a \,\forall\, q \in \{\,q\,\}. \tag{2}$$

Then the flow κ *(q) at the point* $q \in \{q\}$ *induced by* grad z (q) *equals*

$$\kappa\,(q) \;=\; \alpha\,(\;\text{grad } z\,(q)\,,\,q)\;\;\kappa_m\,(q), \tag{3}$$

where α (p , q) is a scalar function of the none negative scalar argument $x \geq 0$ for each fixed $q \in \{\,q\,\}$ and is characterized by (apart from Φ (q)) heat transmitting properties of the given inhomogeneous continuous medium. The function α is known as the modular function of the given continuous medium.

This, then, is the formulation of the principle.

The method of describing the material relation in inhomogeneous continuous media suggested here is justified by the fact that it, in particular, yields, as can be easily verified, the usual material relations which are characteristic feature of linear continuous media. Suppose, for example, that Φ (q) is a set bounded by a hypersurface (indicatrix) in the form of an ellipsoid

$$\frac{\left(\dot{q}^{\,1}\right)^2}{a_1^{\;2}} \;+\; \dots \;+\; =\; \frac{\left(\dot{q}^{\,n}\right)^2}{a_n^{\;2}} \;=\; 1, \tag{4}$$

where lengths of semi-axes a_i = i = 1, ..., n, possibly depend on q. We next set $\alpha\,(x\,,\,q) \equiv |\,x\,|$. Then a straightforward computation based on formulas (1), (2) and (3) shows that

$$\kappa\,(q) = [\,\text{grad } z\,(q)\,.\,B\,(q)\,]^T, \tag{5}$$

where B (q) is an n x n diagonal matrix whose diagonal, in general, consists of distinct quantities depending on a_1, ..., a_n.

It is known [52] that if even two of the quantities a_1, ..., a_n are unequal, then relation (5) describes a linear continuous medium which is non-isotropic with respect to heat transfer. On top of this, if even one of the quantities a_1,..., a_n depends on $q \in \{ q \}$, then the medium is inhomogeneous as well. If all the quantities (or functions of q) are equal to one another, then the medium will be isotropic.

Let us carry out a detailed discussion of the general relation (3). It is clear from (3) that the dependence of the flow κ (q) on grad z (q) is nonlinear. It consists of two multipliers, one is the vector κ_m (q) and the other the scalar α (| grad z (q) |, q). The vector multiplier κ_m (q) stipulates, if one may so express, nonlinear dependence of K (q) on grad z (q), with respect to direction.

Note that (2) obviously implies that κ_m (q) is a positive homogeneous function of degree zero in the variable grad z (q). This means that multiplication of grad z (q) by a number $\lambda > 0$ in (2) does not affect the magnitude of κ_m (q). Therefore, in particular, relations (1) and (2) can be written in the form (with $\alpha \neq 0$)

$$\max_{\kappa \in \Phi(q)} \left[\frac{grad\ z\ (q)}{\alpha\ (\ grad\ z\ (q),q\ )}\ \kappa \right] = \frac{grad\ z\ (q)}{\alpha\ (\ grad\ z\ (q),q\ )}\ \kappa_m\ (q), \qquad (6)$$

$$\kappa_m\ (q) = \arg\max_{\kappa \in \Phi(q)} \left[\frac{grad\ z\ (q)}{\alpha\ (\ grad\ z\ (q),q\ )}\ \kappa \right]. \qquad (7)$$

Formulae (6) and (7) enable us to interpret κ_m (q) as a *flow vector induced by the unit (normalized) temperature gradient*.

The scalar multiplier α (grad z (q) , q) stipulates nonlinear dependence of κ (q) on grad z (q) with respect to magnitudes of these vectors. This can be expressed by saying that λ times ($\lambda > 0$) magnification in | grad z (q) | does not lead, in general, to magnification in | κ (q) | by the same amount. The case when α (p , q), p > 0, represents a proportional dependence is an exception.

We can thus say that relation (3) represents "divided" nonlinear dependence with two multipliers, one of which is responsible for nonlinearity with respect to direction while the other for nonlinearity with respect to magnitude.

The introduction of the modular function α in (3) may seem to be far-fetched. But this is not so. As evidenced by experience and theory [30, 64, 99], there are media, natural as well as artificial, where λ times ($\lambda > 0$) magnification in the vector grad z (q), with no change in its direction, does not lead to a magnification in | κ (q) | by the same amount. For example, there are media where the electric current is not proportional to electric intensity, that is, Ohm's law, a *linear material relation*, is violated. The presence of this kind of nonlinear effects justifies the introduction of the modular function α in Eq. (3).

We now define H (p , q) as the support function of the set Φ (q) (Sec. 8) :

$$H\ (p,q) = \max_{\kappa \in \Phi(q)} p\kappa. \qquad (8)$$

Then formulas (1), (2) and (3) can be written, respectively, in the form

$$H(p,q) = p\kappa_m(q), \tag{9}$$

$$\kappa_m(q) = \underset{\kappa \in \Phi(q)}{\arg\max} \; H(p,q) = \frac{\partial H}{\partial p}(p,q), \tag{10}$$

$$\kappa(q) = \alpha(p,q) = \frac{\partial H}{\partial p}(p,q), \tag{11}$$

where $p = \partial z(q)/\partial q = \operatorname{grad} z(q)$. Formula (11) describes the *nonlinear material relation* in terms of $H(p,q)$.

A physical significance of $H(p,q)$ can be easily explained. Indeed, since $H(p,q)$ is a positive homogeneous function of degree one (Sec.8), we have from (10)

$$p\,\kappa_m(q) = p\frac{\partial H}{\partial p}(p,q) = H(p,q). \tag{12}$$

But $p\,\kappa_m(q)$ gives the amount of heat that flows in a unit time across a unit area, passing through the point q, in the direction of $p = \operatorname{grad} z(q)$ under the action of the unit (normalized) gradient

$$\frac{p}{|p|} = \frac{\operatorname{grad} z(q)}{|\operatorname{grad} z(q)|}.$$

As shown by formula (12), this amount exactly equals $H(p,q)$, where $p = \operatorname{grad} z(q)$. This, then, is the significance of the function $H(p,q)$ (Fig. 28.1).

Likewise, we have from formula (11)

$$p\kappa(q) = (|p|,q)\,p\frac{\partial H}{\partial p}(p,q) = (|p|,q)H(p,q). \tag{13}$$

Formula (13) yields the amount of flow through a unit area, passing through q, in the direction of $p = \operatorname{grad} z(q)$ under the action of $\operatorname{grad} z(q)$.

Now it is an easy task to express the Laplace operator $\Delta z(q)$ in the given inhomogeneous continuous medium in terms of $H(p,q)$ and the modular function $\alpha(|p|,q)$. Indeed, we have, by definition,

$$\Delta z(q) = \operatorname{div} \kappa(qb). \tag{14}$$

Substituting into (14) the expression (11) for $\kappa(q)$, we obtain the desired expression

$$\Delta z(q) = \operatorname{div}\left[\alpha(p,q)\frac{\partial H}{\partial p}(p,q)\right], \qquad p = \frac{\partial z(q)}{\partial q} = \operatorname{grad} z(q). \tag{15}$$

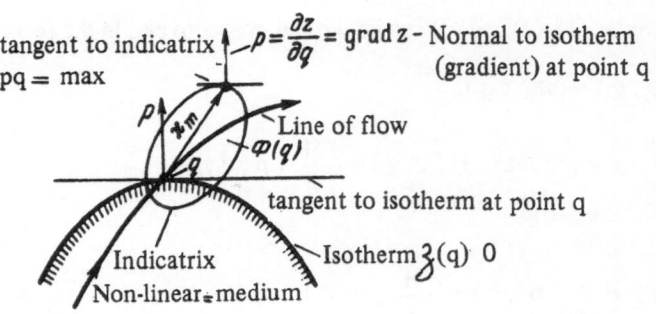

Fig. 28.1

for the Laplace operator in the given inhomogeneous continuous medium. In coordinate form, formula (15) becomes

$$\Delta z(q) = \sum_{K=1}^{n} \frac{\partial}{\partial q^{K}} \left[\alpha(p,q) \frac{\partial H}{\partial p}(p,q) \bigg|_{p = \partial z / \partial q} \right], \tag{16}$$

while in operator form it becomes

$$\Delta = \sum_{K=1}^{n} \frac{\partial}{\partial q^{K}} \left[\alpha(p,q) \frac{\partial H}{\partial p}(p,q) \bigg|_{p = \partial(\cdot)/\partial q} \right],$$

$$= \frac{\partial}{\partial q} \left[\alpha(p,q) \frac{\partial H}{\partial p}(p,q) \bigg|_{p = \partial(\cdot)/\partial q} \right], \tag{17}$$

Heat equations for the unknown temperature distribution z can be written in terms of the Laplace operator. For example, the equation giving steady (statical) distribution of temperature is of the form

$$\Delta z(q) = -\delta(q - q_{o}), \tag{18}$$

where the delta function $\delta(q - q_{o})$ describes the heat source of unit intensity concentrated at the point $q_{o} \in \{q\}$. Eq. (18) can be treated as a limiting case, as t ---> o/o, of the nonstationary heat equation

$$\gamma \frac{\partial z}{\partial t} = \Delta z + g(q), \tag{19}$$

where γ is a coefficient and the function g (q) describes the distribution of stationary (independent of time t) external heat sources in { q }. For the particular case of a point source, g (q) = δ (q $-q_o$). Here the temperature distribution z = z (q , t) depends on t explicitly, and only in the limit, as t ---> oo, the distribution z = z (q), in general, becomes stationary, independent of time.

Now it is an easy task to establish a correspondence between a CDS and the inhomogeneous continuous medium of the type described above. Indeed, let the CDS be governed by the equation

$$\dot{q} = f (q , u), \quad u \in U (q), \quad q \in \{ q \},$$ (20)

or the inclusion

$$\dot{q} \in \Phi (q) \equiv f (q, U (q)),$$ (21)

and let the Hamiltonian of the CDS be of the form

$$H (p , q) = \max_{u \in U (q)} p f (q , u).$$ (22)

We identify the set Φ (q) of the given CDS with the set of admissible flow directions of the continuous medium. In this way, the velocity $\dot{q}$ of CDS (21) is identified with the vector K_m (2). And the Hamiltonian H (p , q) of the CDS is identified with the support function of the set of admissible flow directions. But for the given CDS (20) or (21) the choice of the modular function α remains at our disposal. By choosing a particular modular function α for the given CDS (20) or (21), we associate with this CDS the Laplace operator, defined by formulas (16), (17), of the given continuous medium. This is how a given CDS is associated with a continuous medium and its Laplace operator.

It should be observed that the freedom of choice of the modular function α in such a correspondence proves to be very useful in solving the following inverse problem : Given a continuous medium defined by a partial differential operator, it is required to find the corresponding CDS given by (20), (21). Let us examine the inverse problem in some detail. Let there be given a differential expression (operator) ψ ($\partial^2 z/\partial q^2$, $\partial z/\partial q$, q), where ψ is a given function of the variables $\partial^2 z/\partial q^2$, $\partial z/\partial q$, q. Such an operator can occur in describing some processes in a given continuous medium. It is required to find a CDS governed by (20), (21), (22) (for example, to find H (p, q) to which there corresponds a continuous medium with Laplace operator in the form of ψ.

It is clear that the solution of the inverse problem leads of solving the functional equation

$$\psi \left(\frac{\partial^2 z}{\partial q^2}, \frac{\partial z}{\partial q}, q \right) = \Delta z (q),$$ (22 a)

where Δ z (q) is given by (16) (or (17)). Eq. (22 a) contains two unknown functions, H (p, q) and α (p , q). Of these, H (p, q) must be a positive homogeneous function of degree one in p and α must be a positive function. If such solutions of Eq. (22 a) exist and can be determined, then to the given continuous medium described by ψ there corresponds a CDS of the form

$$\dot{q} = \frac{\partial H}{\partial p}(p, q),$$ (23)

where the vector p plays the role of a control. It is clear that in solving Eq. (22 a), the freedom of choice in α helps find H (p, q) which must satisfy the homogeneity condition.

Let us consider an example on solving the inverse problem. Let

$$\psi \left(\frac{\partial^2 z}{\partial q^2}, \frac{\partial z}{\partial q}, q \right) \equiv \sum_{K=1}^{n} \frac{\partial^2 z}{(\partial q^K)^2}$$ (24)

It is easy to see that in the present case ψ can be identically expressed in the form

$$\psi \equiv \Delta z \equiv \frac{\partial}{\partial z} \left(\frac{\partial z}{\partial q} \right) \equiv \text{div} \frac{\partial z}{\partial q} \equiv \text{div} \left[\left| \frac{\partial z}{\partial q} \right| \cdot \frac{\frac{\partial z}{\partial q}}{\left| \frac{\partial z}{\partial q} \right|} \right] \equiv$$

$$\equiv \text{div} \left[|p| \cdot \frac{p}{|p|} \right], \, p = \frac{\partial z}{\partial q}.$$ (25)

We can easily identify from (25) the desired functions α (| p |, q) and H (p, q) with the required properties. We thus have

$$\alpha(p, q) \equiv |p|, \frac{\partial H}{\partial p} \equiv \frac{p}{|p|}, H(p, q) \equiv |p|.$$ (26)

In this way, the required CDS which corresponds to the continuous medium with operator (24) has an equation of the form

$$\dot{q} = \frac{\partial H}{\partial p} = \frac{p}{|p|},$$ (27)

where p plays the role of a control.

Since H (p, q) is the support function of the convex set f (q, U (q)) $\equiv \Phi$ (q), it is easy to recover the set itself : it is a disc of radius 1 with centre O° (q). Therefore the equation of the CDS, in terms of control u, becomes

$$\dot{q} = u, \quad u \in U = \{ u : |u| \leq 1 \}.$$ (28)

CDS (28) is very simple ; for it the entire state space $\{ q \} = R_n$ is the domain of uncons trained motions, since

$$\min_{|p|=1} H(p,q) = \min_{|p|=1} |p| = 1 > 0 \qquad \forall q \in \{q\}, \qquad (29)$$

in view of (9.2).

To conclude the pressent section, we discuss the significance of the results obtained. It should be remarked that the correspondence (one-to-one correspondence) between CDSs and processes in continuous media, established above, can prove to be useful in studying one object with the aid of the other. Problems formulated for, say, continuous media can also be formulated as corresponding problems for CDSs, and vice-versa. Thus, for example, the controllability problem of a CDS can be solved in terms of heat conduction. Indeed, assume that to the given CDS there corresponds a continuous medium with the Laplace operator $\Delta z (q)$ of the form (16), (17). Suppose that it has been proved that Eq. (18), which establishes a heat conduction process, has a nonvanishing solution $z(q)$ for all $q \in \{q\}$. Then it can be concluded that the given CDS is controllable in $\{ q \}$. If at some point $z(q) = 0$, then it can be concluded that the CDS is uncontrollable.

The inverse problem of investigating a continuous medium with the aid of the corresponding CDS is also very prospective. As a matter of fact, in recent years there have arisen many problems in various fields of science and technology where one has to deal with very complex continuous media. Equally complex are the partial differential equations governing processes in these media. Therefore the possibility of reformulating problems for continuous media in terms of problems for the corresponding CDSs can prove to be very reassuring.

Since real physical continuous media occupy the space $\{ q \}$ of small dimension $n = 1$, 2, 3, the corresponding CDS is also of the same small dimension n. And for a CDS of a small dimension a very effective method of investigation is the method of phase portrait of the CDS, treated in the present book.

Finally, we consider one more possibility of assigning to the function $\mathcal{H}(p, q)$, which is not necessarily homogeneous, a continuous medium with the Laplace operator $\Delta z (q)$. The function $\mathcal{H}(p, q)$ can be interpreted as the Hamiltonian of a mechanical system. To accomplish this, the flow $\kappa(q)$ is identified with $\partial\mathcal{H}(p, q)/\partial p$, with $p = \text{grad } z$:

$$\kappa(q) = \frac{\partial \mathcal{H}}{\partial p}(\text{grad } z(q), q). \qquad (30)$$

Then

$$\Delta z(q) = \text{div } \kappa(q) = \text{div}_q \frac{\partial \mathcal{H}}{\partial p}(\text{grad } z(q), q). \qquad (31)$$

which yields

$$\Delta z\,(q) = t\,r\left[\frac{\partial^2 \mathcal{H}}{\partial p^2}(\text{grad } z\,(q), q)\frac{\partial^2 z}{\partial q^2} + \frac{\partial^2 \mathcal{H}}{\partial p \partial q}\,(\text{grad } z\,(q), q)\right],\tag{32}$$

where $t\,r\,[\cdot]$ denotes the trace of the matrix present in the square bracket. In particular, for $\mathcal{H}\,(p, q) = \frac{p^2}{2}$, we obtain the classical Laplacian

$$\Delta z = \sum_{i=1}^{n} \frac{\partial^2 z}{(q^i)^2}.$$

29

CDS and Finsler Metric

A Finsler metric is a generalization of a Riemannian metric. A Riemannian metric can be treated as a particular case of a Finsler metric when its indicatrix at every point of the phase space is a nondegenerate central hypersurface of second order having the equation

$$\sum_{\beta=1}^{n} \sum_{\alpha=1}^{n} g_{\alpha\beta} \dot{q}^{\alpha} \dot{q}^{\beta} = 1. \tag{1}$$

the coefficients $g_{\alpha\beta}$ in general, depend on q. In particular, (1) is the equation of an ellipsoid

$$\frac{(\dot{q}^1)^2}{\dot{q}_1^2} + \dots + \frac{(\dot{q}^n)^2}{\dot{q}_n^2} = 1, \tag{2}$$

where the lengths of the semi-axes a_i of the ellipsoid may depend on q. All the vectors have a definite length if the quadratic form in (1) is positive definite. However, this form is not always positive definite. For instance, in the theory of relativity this form is not positive definite, and this corresponds to the fact that the cone of admissible velocities (light cone) K (q) does not coincide with the entire space $T_q \{ q \}$. In this case, only vectors belonging to K (q) have a definite length. In Finsler geometry, K (q) is, as a rule, of full dimension n. As we have seen, for a CDS the dimension of this cone can be less than n.

A Finsler metric is defined as follows [95]. An indicatrix, a hypersurface is defined in the tangent space $T_q \{ q \}$ such that every ray originating from the point $\dot{O}$ (q) intersects this hypersurface in at most one point and the tangent plane to this hypersurface does not pass through $\dot{O}(q)$. Let η be a prescribed vector in $T_q \{ q \}$. In $T_q \{ q \}$, draw a ray $\dot{O} = \dot{O}$ (q), from collinear with η, up to the point of intersection with the indicatrix. This ray intersects, the indicatrix in one point R. The vector $\dot{OR}$, joining $\dot{O}$ and the indicatrix, is taken as a unit vector in the given direction η. Thus the vectors η and $\dot{OR}$ are collinear, and the ratio of their magnitudes $|\eta| / |\dot{O}R|$ is called the *length of the vector in the Finsler metric*. New the indicatrix can be treated as a geometrical figure consisting of endpoints of unit vectors originating from the point $\dot{O}(q) \in T_q \{ q \}$.

It should be observed that in Finsler geometry, an indicatrix is always a hypersurface in $T_q \{ q \}$. The vectors η not intersecting the indicatrix are regarded as nonmeasurable. However, for a CDS the indicatrix is not always a hypersurface (Sec.6), and hence it cannot be always identified with the indicatrix of the Finsler metric. The indicatrix of a CDS often lies in linear subspace L_m whose dimension $m < n$. Thus in order for us to be able to use metric notions for a CDS with an arbitrary indicatrix (Sec.6), it is necessary to extend the meaning of an indicatrix in Finsler geometry by including in its rank the "relative" hypersurfaces lying in the subspace L^m $T_q \{ q \}$ with $m < n$.

What is more, the indicatrix of a CDS bounds a convex set which often does not contain the point $\dot{O}(q)$ of $T_q \{ q \}$. Therefore a ray, say, collinear with the given vector η may intersect the indicatrix in more than one point. In such a case, we can speak of the *maximum length* $|\eta| / |\dot{O}R_o|$ and the *minimum length* $|\eta| / |\dot{O}R_1|$ of the vector η (Fig. 29.1). The portion AD of the indicatrix corresponds to the minimum length of η while BC to the maximum length. To these two portions there correspond two Lagrangians (or two branches of the same Lagrangian) L_0 and L_1, which can be associated with the given CDS, with indicatrix having the aforementioned structure [100]. If we talk

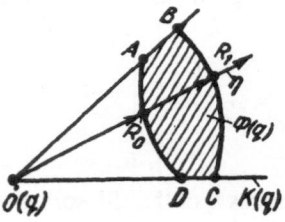

Fig. 29.1

of an optical analogy of the CDS (Sec.30), then L_0 corresponds to the minimum velocity of propagation of perturbation front while L^1 to the maximum velocity for the given CDS. This situation occurs when the indicatrix is given, for example, by a translated $(n - 1)$-dimensional sphere (hypersurface) having equation

$$\sigma(\dot{q}, q) \, |\dot{q} - a(q)| - r(q) = 0, \quad |a(q)| > r(q) \geq 0,$$

where $r(q)$ is the radius of the sphere and $a(q)$ is the shift vector of the sphere's centre from the origin of $T_q \{ q \}$.

Thus trajectories of the given CDS in its state space obtained from the canonical equations corresponding to the CDS can be treated as geodesic curves, that is, curves having locally minimum lengths between any two points in the corresponding Finsler (in particular, Riemannian) metric. Such geodesic curves are minimal time trajectories.

Optical Analogue of CDS

An analogy between a CDS and the process of propagation of perturbation of, for example, light in a continuous medium is carried out in a simple and natural manner. Here we have in view the geometrical theory of perturbation ; in particular, geometrical optics [9, 19]. From the mathematical point of view, this theory is very lucid, and a short account of it can be found in Supplement 1 of [35]. Thus in order for us to carry out the analogy between processes described by geometrical optics and CDS, we have only to make the following identification : the indicatrix of the velocities of perturbation (of light) is identified with the indicatrix of admissible velocities of the given CDS. Then the optical analogy of the CDS becomes plain, and it can be developed in exactly the same way as done in [35].

Here we confine ourselves to the derivation of the associated Lagrangian L $(\dot{q}, q)$. Recall that according to the Fermat principle, the perturbation (light) between two points q_0 and q_1 is propagated along the trajectory (ray) $q(t)$, $q(0) = q_0$, $q(T) = q_1$, in the state space $\{q\}$ such that the action integral (optical length of path)

$$
I = \int_{q_0}^{q_1} \frac{|dq|}{\vartheta(\eta, q)} = \int_0^T L(\dot{q}, q) \, dt \tag{1}
$$

attains its extremal (minimum) value. Here $1/v(\eta, q)$ denotes the refractive index and $v(\eta, q)$ the magnitude of the velocity of light at the point q in the direction of η. By comparing the two integrals in (1) and noting that $|dq| = |\dot{q}|$ we have

$$
\frac{|\dot{q}|}{\vartheta(\eta, q)} \, dt = L(\dot{q}, q) \, dt.
$$

Hence the desired Lagrangian is of the form

$$L(\dot{q}, q) = \frac{|\dot{q}|}{\vartheta(\eta, q)}. \tag{2}$$

If the indicatrix of the CDS is given by the equation (Sec.5)

$$\sigma(\dot{q}, q) = 0, \tag{3}$$

then by expressing this equation in the form $\sigma(|\dot{q}| \eta, q) = 0$, where $\eta = \dot{q}/|\dot{q}|$ is a unit vector in the direction of $\dot{q}$, and then solving this equation for $|\dot{q}|$, we obtain

$$|\dot{q}| = \vartheta(\eta, q) = \vartheta\left(\frac{\dot{q}}{|\dot{q}|}, q\right). \tag{4}$$

Thus the desired Lagrangian is in the form of a positive and homogeneous function of degree one in $\dot{q}$:

$$L(\dot{q}, q) = \frac{|\dot{q}|}{\vartheta\left(\dfrac{\dot{q}}{|\dot{q}|}, q\right)}, \tag{5}$$

where $v(\dot{q}/|\dot{q}|, \dot{q})$ is given by (4) and plays the role of the function $(\dot{q}, q)$. The equation of the indicatrix then assumes the form

$$L(\dot{q}, q) = 1. \tag{6}$$

As noted in Sec. 29, if the ray originating from $\dot{O}(q)$ in the direction of η intersects the indicatrix in more than one point, then Eq. (3) in $|\dot{q}|$ has more than one solution. Therefore we can speak of the maximum perturbation velocity $v_1(\eta, q)$ in the direction of η as well as the minimum velocity $v_0(\eta, q)$ in the same direction. In this case, we can speak of rapid and slow perturbation fronts with which there are associated the two Lagrangians

$$L_1(q, q) = \frac{|q|}{\vartheta_1\left(\dfrac{q}{|q|}, q\right)}, \quad L_0(q, q) = \frac{|q|}{\vartheta_0\left(\dfrac{q}{|q|}, q\right)}. \tag{7}$$

It is also clear from (5) that if η or $(\dot{q})$ belongs to the cone $K(q)$ of admissible directions, then L is a definite finite quantity. If η or $(\dot{q})$ does not lie in $K(q)$, then L should be treated as an undefined quantity or as an infinitely large positive quantity. The source of light (the source of perturbation) located at the point q' will illuminate only those points of $\{q\}$ which are reachable from q' under the action of the admissible controls of the CDS.

31

Correspondence between CDSs and Mechanical Systems

Since there are well developed methods for investigating Hamiltonian mechanical system, the fact that CDSs can be associated with mechanical system can prove to be useful in studying CDSs. In particular, to be very useful in the study of CDSs prove the tools of canonical transformations, generating functions and several other effective methods of analytical mechanics.

Consider the CDS

$$\dot{q} = f(q, u), \quad u \in U(q). \tag{1}$$

As we have seen (Sec. 14), with this CDS we can associate a unique uncontrollable Hamiltonian mechanical system with a Hamiltonian of the form

$$H(p, q) = \sup_{u\,U(q)} \quad (\max) pf(q, u). \tag{2}$$

which is a positive homogeneous function of degree one in p. Conversely, suppose we have an uncontrollable mechanical system with the Hamiltonian H (p, q) which is a convex, positive homogeneous function of degree one in p. Then with this system we can associate a CDS of the form

$$\dot{q} = \frac{\partial H}{\partial p}(p, u), \tag{3}$$

where the parameter p plays the role of a control with no condition whatsoever imposed on it. Since $\partial H/\partial p$ in (3) is a homogeneous function of degree zero in p, that is, in reality it depends only on p/ | p |, the control p can be assumed to vary only on, for example, the sphere | p | = 1.

It should be observed that now the state space { q } of CDS (1) and (3) becomes a configuration (or coordinate) space with respect to the uncontrollable mechanical system with the Hamiltonian (2). And, conversely, the configuration space of the uncontrollable mechanical system with the Hamiltonian H (p, q) becomes the state space (phase space) of the associated CDS (3) . For CDS (3), the control space (the set of admissible controls) coincides with the impulse space { p } of the same mechanical system. If we now try to associate with CDS (3) a Hamiltonian system having a positive homogeneous Hamiltonian function of degree one, we again obtain the given Hamiltonian H (p, q). Indeed, corresponding to CDS (3) a positive homogeneous Hamiltonian function of degree one is defined by the formula

$$\mathcal{H}(\pi, q) = \sup_{p} (\max) \; \pi \; \frac{\partial H}{\partial p} \; (p, q). \tag{4}$$

But $\dot{q} = \partial H / \partial p$ is just what yields that value of $\dot{q}$ where the upper bound (maximum) of the quantity $p \dot{q}$ is attained. Thus the upper bound (or maximum) in (4) is attained at $p = \pi$. Consequently, $\mathcal{H}(\pi, q) = \frac{\partial H}{\partial p} \; \pi \; (\pi, q) = H (\pi, q)$, as required.

In what follows, two positive homogeneous Hamiltonian functions of degree one H_1 (p, q) and H_2 (p, q) will be called *equivalent* if the sets of points of { p } described by the equations H_1 (p, q) = 0 and H_2 (p, q) = 0 for any $q \in$ { q } are either both empty or both describe the same cone $\overline{K}$ (q). The set of all positive and homogeneous of degree one hamiltonians H (p, q) which are equivalent constitutes an equivalence class defined by the field of cones $\overline{K}$ (q). It is assumed here that for some $q \in$ { q } the cone may be empty.

Thus, assuming that the space { $\dot{q}$ } is conjugate of the empty cone $\overline{K}$ (q), we can establish a one-to-one correspondence

$$\dot{q} \in K (q) \Leftrightarrow \overline{K} (q) \; H (p, q) \le 0, \qquad \forall \, q \in \{ q \}, \tag{A}$$

between the equivalence classes of generalized trajectories of CDS determined by the field of cones K (q) (it is assumed that K (q) coincides with { $\dot{q}$ }) and the equivalence classes of positive and homogeneous of degree one Hamiltonians H (p, q) characterized by the field of cones $\overline{K}$ (q) (this includes the empty cones too).

This correspondence is, in fact, a consequence of the one-to-one correspondence between the cones K (q) and $\overline{K}$ (q). That is, a consequence of the relation $\overline{\overline{K}} (q) = K (q)$, $\overline{K} (q) = K (q)$, which shows that the operation of defining conjugate of convex cones is an involution.

We shall now discuss how to establish the correspondence between a mechanical system and a CDS. If the mechanical system with n degrees of freedom is defined by a Hamiltonian H (p, q) which does not constitute a positive homogeneous function of degree one in p, then we first "homogenize" it by the formula

$$H_{+1}\ (p_0, p, q) = |\,p_0\,|\ H\left(\frac{p}{|\,p_0\,|}, q\right), \tag{5}$$

where p_0 is the new coordinate of the impulse corresponding to the new coordinate of the state q°. Function (5) is taken as the Hamiltonian for the desired CDS. Then the associated CDS is of order $n+1$ and is governed by the equations

$$\dot{q}^0 = \frac{\partial H_{+1}}{\partial p_0} = \text{sign}\,(-p_0)\left[H\left(\left|\frac{p}{|\,p_0\,|}, q\right) - \frac{p}{p_0}\frac{\partial H}{\partial p}\left(\frac{p}{|\,p_0\,|}, q\right)\right], \tag{6}$$

$$\dot{q} = \frac{\partial H_{+1}}{\partial p} = \frac{\partial H}{\partial p}\left(\frac{p}{|\,p_0\,|}, q\right). \tag{7}$$

In CDS (6), (7), the role of the control is played by the vector (p_0, p) on which no condition whatsoever is imposed. The adjoint system is of the form

$$\dot{p}_0 = -\frac{\partial H_{+1}}{\partial q^0} = 0, \tag{8}$$

$$\dot{p} = -\frac{\partial H_{+1}}{\partial q} = -|\,p_0\,|\frac{\partial H}{\partial q}\left(\frac{p}{|\,p_0\,|}, q\right), \tag{9}$$

In order that the system (7), (9) coincides with the canonical equations of the original mechanical system it is required that $|\,P_0\,| = 1$. Then the expression enclosed by the square bracket coincides with the Lagrangian $L\,(\dot{q}, q)$, corresponding to $H\,(\,p, q\,)$. Further, in order to assign q^0 the sense of a mechanical action, we set $p_0 = -1$.

Then (6) assumes the form

$$\dot{q}^0 = L\,(\dot{q}, q), \tag{10}$$

and from (10) we have

$$\sigma\,(\dot{q}^0, \dot{q}, q) \equiv \dot{q}^0 - L\,(\dot{q}, q) = 0. \tag{11}$$

Eq. (11) represents the indicatrix of the $(n+1)$ th-order CDS with state vector (q^0, q). And this is the CDS that is associated with the original mechanical system with the Hamiltonian $H\,(\,p, q\,)$ and the Lagrangian $L\,(\dot{q}, q)$. On account of (10), we can write the CDS in the form of a normal system of differential equations

$$\overset{0}{\dot{q}} = L (u , q),\tag{12}$$

$$\dot{q} = u,\tag{13}$$

where no condition is imposed on the control $u = (u^1 , \dots , u^n)$.

In our view it is pertinent to consider, in the present section, another question which in a definite sense touches the problems regarding relationship between CDSs and uncontrollable mechanical systems which are subjected to additional integrability or non-integrability conditions.

It is known [78] that in many cases (for example, when the external forces acting on the given system constitute a potential system of forces) the equations of motion of a holonomic system which is subjected to ideal constraints

$$f (q , t) = (f_i (q , t)) = 0,\tag{14}$$

or to constraints, equivalent to (14),

$$\pi (q , t) \, \dot{q} + \frac{\partial f}{\partial t} = 0,\tag{15}$$

$$\pi (q, t) = (\pi_{ij} \, q , t)) \equiv \frac{\partial f}{\partial q} = \left(\frac{\partial f_i}{\partial q_j} \right),\tag{16}$$

are of the form

$$\frac{d}{dt} \left(\frac{\partial L}{\partial \dot{q}} \right) - \frac{\partial L}{\partial q} = \lambda \, \pi.\tag{17}$$

The index i here, and in the sequal, runs through 1 to m and j from 1 to n, m < n; $q = (q_{ij})$ denotes the generalized coordinates $\dot{q}$, the generalized velocity, $\pi = \pi (q, t)$ is an m x n matrix with full rank, $L = L (\dot{q}, q, t)$ is the Lagrangian of the system without constraints and λ is an m-vector. It is assumed that all the derivatives required exist and are continuous, while the initial point $q (t_0)$ and the initial velocity $q (t_0)$ are assumed to satisfy (14), (15).

Assume that the Hamiltonian $H (p, q, t)$ is in a one-to-one correspondence with $L (\dot{q}, q, t)$ by means of the Legendre transformation. Then Eqs. (15), (17) are transformed to

$$\dot{q} = \frac{\partial H}{\partial p} \, (p, q, t),\tag{18}$$

$$\dot{p} = - \frac{\partial H}{\partial q} \, (p, q, t) + \lambda \, \pi (q, t).\tag{19}$$

which is close to the canonical form. Eqs. (18), (19) are *not exactly* canonical due to presence of the member $\lambda \, \pi$ in (19). The system of Eqs. (15), (17) or of Eqs. (18), (19), (15) constitutes a complete system in the unknown functions q (t), λ (t) or q (t), p (t), λ (t).

   We shall show that there exists a function $\mathcal{H} = \mathcal{H}(\, \mathcal{P}, q, t \,)$, where $\mathcal{P} = (\, \mathcal{P}_j)$ is the new impulse which is defined by the relation (maximum condition with respect to $\dot{q}$)

$$\mathcal{H}(\, \wp, q, t \,) \; = \; \max \; \left(- L \, (\, \dot{q}, q, t \,) \; + \; \wp \, \dot{q} \, | \, \pi \, \dot{q} + \frac{\partial f}{\partial t} \, = \, 0 \right) \tag{20}$$

and is such that the equations of motion of the holonomic system in question are given in the exact canonical form

$$\dot{q} \; = \; \frac{\partial \mathcal{H}}{\partial \wp} \; (\, \wp, q, t \,) \tag{21}$$

$$\dot{\wp} \; = \; - \frac{\partial \mathcal{H}}{\partial q} \; (\, \wp, q, t \,) \tag{22}$$

That is, $\mathcal{H}(\, \wp, q, t \,)$ is the exact Hamiltonian of the holonomic system subject to the constraints (14), (15). We shall demonstrate that relations (21), (22) imply the relations (15), (18), (19), and hence (15), (17) as well. To do this, we first perform the required maximization in (20). Maximum conditions (20) with repect to $\dot{q}$ imply that for fixed $\wp$, q, t there exists a vector $\mu = (\, \mu_i)$ such that the function

$$- L \, (\, \dot{q}, q, t \,) \; + \; \wp \, \dot{q} + \mu \left(\pi \, \dot{q} + \frac{\partial f}{\partial t} \right) \tag{22 a}$$

has an extremum with respect to $\dot{q}$ satisfying the equation

$$- \frac{\partial L}{\partial \dot{q}} \; + \; \wp \; \mu \pi \; = \; 0 \tag{23}$$

or

$$\wp + \mu \, \pi \; = \; \frac{\partial L}{\partial \dot{q}} \tag{24}$$

Employing the notation

we have, from (24),

$$p \; = \; \wp \; + \; \mu \pi, \tag{25}$$

$$p \; = \frac{\partial L}{\partial \dot{q}} \; (\, \dot{q}, q, t \,). \tag{26}$$

Since H and L are in a one-to-one correspondence, we can solve (26) for $\dot{q}$ and obtain

$$\dot{q} = \frac{\partial H}{\partial p} \ (p, q, t).$$ (27)

By virtue of (25), Eq. (27) becomes

$$\dot{q} = \frac{\partial H}{\partial p} \ (\wp + \mu\pi, q, t).$$ (28)

Taking into account (20), we now determine μ by the condition that constraint (15) is to be satisfied ; this yields

$$\pi_{i\beta} \ (q) \ \frac{\partial H}{\partial p_\beta} \ (\wp + \mu\pi, q, t) + \frac{\partial f_i}{\partial t} = 0.$$ (29)

It is assumed here, and in the sequel, that the summation is performed over α from 1 to m and over β from 1 to n. Clearly, the desired function μ is a function of $\wp$, q, t, that is, $\mu = \mu \ (\wp, q, t)$. In what follows, we assume that Eq. (29) is uniquely solvable for μ.

It should be observed that if the known function $\mu = \mu \ (\wp, q, t)$ is substituted into (29) and (28), Eq. (29) becomes an identity. And Eq. (28) (in view of identity (29)) automatically ensures that constraints (15), (16) are satisfied for any $\wp$, q, t.

Further, by substituting (28) into (22a) with $\mu = \mu \ (\wp, q, t)$, we obtain the explicit expression

$$\mathcal{H} \ (\wp, q, t) = -L\left(\frac{\partial H}{\partial p} (\wp + \mu\pi, q, t), q, t \right) + \wp \frac{\partial H}{\partial p} (\wp + \mu\pi, q, t).$$ (30)

for the desired Hamiltonian. We add and subtract the expression

$$\mu\pi \frac{\partial H}{\partial p} \ (\wp + \mu\pi, q, t)$$

on the right hand side of (30). Then (30) becomes

$$\mathcal{H} \ (\wp, q, t) = -L\left(\frac{\partial H}{\partial p} (\wp + \mu\pi, q, t), q, t \right) +$$

$$+ \ (\wp + \mu\pi) \frac{\partial H}{\partial p} \ (\wp + \mu\pi, q, t) - \mu\pi \frac{\partial H}{\partial p} \ (\wp + \mu\pi, q, t)$$ (31)

By virtue of the identity

$$H \ (p, q, t) = -L\left(\frac{\partial H}{\partial p} \ (p, q, t), q, t \right) + p \frac{\partial H}{\partial p} (p, q, t)$$ (32)

where $p = \wp + \mu\bar{\pi}$ is substituted, from (31) we obtain

$$\mathcal{H}(\wp, q, t) = H(\wp + \mu(\wp, q, t)\pi(q, t), q, t) -$$

$$- \mu(\wp, q, t)\pi(q, t)\frac{\partial H}{\partial p}(\wp + \mu(\wp, q, t)\pi(q, t), q, t). \qquad (33)$$

Identity (29) enables us to write (33) in the alternative form

$$\mathcal{H}(\wp, q, t) = H(\wp + \mu(\wp, q, t)\pi(q, t), q, t) + \mu(\wp, q, t)\frac{\partial f}{\partial t}(q, t). \qquad (34)$$

Thus the solution of the maximization problem (20) is given by (28) and (34), where $\mu = \mu(\wp, q, t)$ is determined from solution of (29). It should be observed that $\mathcal{H}$ and H naturally coincide, as is clear from (20) and (34), if the constraints (14)–(16) are absent.

It will be shown by a direct computation that Eq. (21) implies Eq. (18) and Eq. (22) implies Eq. (19). We have

$$\dot{q} = \frac{\partial \mathcal{H}}{\partial \wp} = \frac{\partial}{\partial \wp}\left[H(\wp + \mu\pi, q, t,) + \mu \frac{\partial f}{\partial t}\right] =$$

$$= \frac{\partial H}{\partial p}(\wp + \mu\pi, q, t) + \frac{\partial \mu}{\partial \wp}\pi\frac{\partial H}{\partial p}(\wp + \mu\pi, q, t) + \frac{\partial \mu}{\partial \wp}\cdot\frac{\partial f}{\partial t} =$$

$$= \frac{\partial H}{\partial p}(\wp + \mu\pi, q, t) - \frac{\partial \mu}{\partial \wp}\cdot\frac{\partial f}{\partial t} + \frac{\partial \mu}{\partial \wp}\cdot\frac{\partial f}{\partial t} =$$

$$= \frac{\partial H}{\partial p}(\wp + \mu\pi, q, t) = \frac{\partial H}{\partial p}(p, q, t). \qquad (35)$$

In arriving at (35), we have successively used identity (29) and formula (25) for inverting the new impulse $\wp$ in terms of the original impulse p. Thus (21) implies (18).

It remains to be shown that Eq. (22) implies Eq. (19). Indeed, the left side of (22), by virtue of (25), yields

$$\dot{\wp}_j = \frac{d}{dt}(p_j - \mu_\alpha \pi_{\alpha j}) = \dot{p}_j - \dot{\mu}_\alpha \pi_{\alpha j} - \mu_\alpha\left(\frac{\partial \pi_{2j}}{\partial q_\beta}\cdot\dot{q}_\beta - \frac{\partial \pi_{\alpha j}}{\partial t}\right).$$

$$= \dot{p}_j - \dot{\mu}_\alpha \pi_{\alpha j} - \mu_\alpha\left(\frac{\partial \pi_{\alpha j}}{\partial q_\beta}\frac{\partial H}{\partial p_\beta} - \frac{\partial^2 f_\alpha}{\partial q_j \partial t}\right). \qquad (36)$$

Here formulae (28) and (16) have been used. On the other hand, for the right side of (22) we have

$$-\frac{\partial \mathcal{H}}{\partial q_j} = -\frac{\partial}{\partial q_j}\left[H\left(\wp + \mu\pi, q, t\right) - \mu\frac{\partial f}{\partial t}\right] =$$

$$= -\frac{\partial H}{\partial q_j} - \frac{\partial\left(\mu_\alpha \pi_{\alpha\beta}\right)}{\partial q_j}\frac{\partial H}{\partial p_\beta} - \frac{\partial}{\partial q_j}\left(\mu_\alpha\frac{\partial f_\alpha}{\partial t}\right) =$$

$$= -\frac{\partial H}{\partial q_j} - \left[\frac{\partial \mu_\alpha}{\partial q_j}\pi_{\alpha\beta} + \mu_\alpha\frac{\partial \pi_{\alpha\beta}}{\partial q_j}\right]\frac{\partial H}{\partial p_\beta} - \frac{\partial \mu_\alpha}{\partial q_j}\frac{\partial f_\alpha}{\partial t} - \mu_\alpha\frac{\partial^2 f_\alpha}{\partial t\,\partial q_j} =$$

$$= -\frac{\partial H}{\partial q_j} - \frac{\partial \mu_\alpha}{\partial q_j}\pi_{\alpha\beta}\frac{\partial H}{\partial p_\beta} - \mu_\alpha\frac{\partial \pi_{\alpha\beta}}{\partial q_j}\frac{\partial H}{\partial p_\beta} - \frac{\partial \mu_\alpha}{\partial q_j}\frac{\partial f_\alpha}{\partial t} - \mu_\alpha\frac{\partial^2 f_\alpha}{\partial t\,\partial q_j} \cdot \quad (37)$$

Since, on account of (29),

$$\pi_{\alpha\beta}\left(\frac{\partial H}{\partial p_\beta}\right) = -\frac{\partial f_\alpha}{\partial t},$$

we find that the second and fourth members in (37) cancel out each other. Thus (37) assumes the form

$$\frac{\partial \mathcal{H}}{\partial q_j} = -\frac{\partial H}{\partial q_j} - \mu_\alpha\frac{\partial \pi_{\alpha\beta}}{\partial q_j}\frac{\partial H}{\partial p_\beta} - \mu_\alpha\frac{\partial^2 t_\alpha}{\partial t\,\partial q_j}. \quad (38)$$

Equating the right sides of (36) and (38) (in view of the original Eq. (22)), we obtain a new equation in which the last members of the left and right hand sides are equal ($\partial^2 f_\alpha/\partial q_j\,\partial t = \partial^2 f_\alpha/\partial t\,\partial q_j$) and are therefore eliminated. Then Eq. (22) becomes

$$\dot{p}_j = -\frac{\partial H}{\partial q_j} + \mu_\alpha\left(\frac{\partial \pi_{\alpha j}}{\partial q_\beta} - \frac{\partial \pi_{\alpha\beta}}{\partial q_j}\right)\frac{\partial H}{\partial p_\beta} + \dot{\mu}_\alpha\,\pi_{\alpha j} \quad (39)$$

For arbitrary π_{ij} (q, t) the "vertical" quantitites in the parentheses in (39) do not vanish. But under the assumption that constraints (14), (15) are holonomic, that is, under the assumption that condition (16) is satisfied, these quantities vanish identically, Indeed, in view of (16), we have

$$\frac{\partial \pi_{\alpha j}}{\partial q_\beta} - \frac{\partial \pi_{\alpha\beta}}{\partial q_j} = \frac{\partial^2 f_\alpha}{\partial q_j\,\partial q_\beta} - \frac{\partial^2 f_\alpha}{\partial q_\beta\,\partial q_j} = 0. \quad (40)$$

If we take into account (40), we find that (39) coincides with (19) provided λ in the latter equation is identified with μ. Since under the assumptions made and notation employed, all the computations are reversible. Eqs. (15), (18), (19) imply Eqs. (21), (22). Thus we have established the equivalence of the system of Eqs. (14)-(17) or of Eqs. (14), (15), (18), (19) with the system of Eqs. (21), (22), (20), (34), where λ (t) = $\dot{\mu}$ (t).

The present question can be investigated from another angle too. It can be easily shown that Eqs. (14)-(17) are equations of extremals for the problem of conditional extremum of the functional $\mathcal{L}$

$$\int_0^T L(\dot{q}, q, t) \, dt \tag{41}$$

subject to the constraint (14). Indeed, it is known that [63a] there exists a function λ (t) such that the extremal q (t) satisfies Euler's equation

$$\frac{d}{dt}\left(\frac{\partial \mathcal{L}}{\partial \dot{q}}\right) - \frac{\partial \mathcal{L}}{\partial q} = 0 \tag{42}$$

and the constraint (14), where

$$\mathcal{L}(\dot{q}, q, t) \equiv \mathcal{L} = L(\dot{q}, q, t) + v(t) \, f(q, t). \tag{43}$$

Substituting (43) into (42), we obtain

$$\frac{d}{dt}\left[\frac{\partial}{\partial \dot{q}}\left(L(\dot{q}, q, t) + v(t)f(q, t)\right)\right] - \frac{\partial}{\partial q}\left[L(\dot{q}, q, t) + v f(q, t)\right] =$$

$$\frac{d}{dt}\left(\frac{\partial L}{\partial \dot{q}}\right) - \frac{\partial L}{\partial q} - v(t)\frac{\partial f}{\partial q} = 0, \tag{44}$$

which coincides with (17).

In the problem of conditional extremum of (41), if the constraint is taken in the form (15), (16), instead of (14), then, as is known, again there exists a function v (t) such that the extremal q (t) satisfies Euler's equation

$$\frac{d}{dt}\left(\frac{\partial \mathcal{L}_1}{\partial \dot{q}}\right) - \frac{\partial \mathcal{L}_1}{\partial q} = 0, \tag{45}$$

and constraints (15), (16), where now $\mathcal{L}_1$ differs from (43) and is given by

$$L_1 (\dot{q}, q, t) = L (\dot{q}, q, t) + v_\alpha \left[\pi_{\alpha\beta} \dot{q}\beta + \frac{\partial f_\alpha}{\partial t} \right]. \tag{46}$$

However, here too substituting (46) into (45), we again obtain Eq. (17). Indeed, taking into account (16), we have

$$\frac{d}{dt} \left[\frac{\partial}{\partial \dot{q}} \left(L (\dot{q}, q, t) + v_\alpha(t) \, \pi_{\alpha\beta} \dot{q}\beta + v_\alpha(t) \, \frac{\partial f_\alpha}{\partial t} \right) \right] -$$

$$= \frac{\partial}{\partial q} \left[L (\dot{q}, q, t) + v_\alpha(t) \, \pi_{\alpha\beta} \dot{q}_\beta + v_\alpha(t) \, \frac{\partial f_\alpha}{\partial t} \right] =$$

$$- \frac{d}{dt} \left(\frac{\partial L}{\partial \dot{q}} \right) + \dot{v}_\alpha \pi_{\alpha\beta} + v_\alpha \frac{\partial \pi_{\alpha\beta}}{\partial q_\beta} \dot{q}_\beta + v \frac{\partial^2 f}{\partial q \partial t} - \frac{\partial L}{\partial q} -$$

$$- v \frac{\partial^2 f}{\partial q^2} \dot{q} - v \frac{\partial^2 f}{\partial t \partial q} = \frac{d}{dt} \left(\frac{\partial L}{\partial \dot{q}} \right) - \frac{\partial L}{\partial q} + \dot{v} \frac{\partial f}{\partial q} = 0, \tag{47}$$

which coincides with (14) if λ is identified with $\dot{v}$.

It can be concluded once again from the above discussion that the principle of extremal action is preserved in the case of nonlinear holonomic constraints or "nonholonomic" integrable constraints [2].

On the other hand, the extremal problem for the functional (41) subject to the conditions (15), (16) can be solved with the aid of the maximum principle [4]. By virtue of this principle, the extremal q (t) satisfies the following Hamiltonian system of equations

$$\dot{q} = \frac{\partial \mathcal{H}}{\partial \wp} (\wp, q, t), \tag{48}$$

$$\dot{\wp} = - \frac{\partial \mathcal{H}}{\partial q} (\wp, q, t), \tag{49}$$

where $\mathcal{H}(\wp, q, t)$ is given by (7). Thus we again obtain a system of hamiltonian Eqs. (48), (49) which is exactly the system (8), (9). However, the derivation of these equations carried out in the beginning of the present section should not be dismissed as unnecessary for it demonstrates the explicit relationship between the old and new impulses p and $\wp$ (see (25)) as well as between the functions H (p, q, t) and $\mathcal{H}(\wp, q, t)$ (see (34)).

We conclude the present sections with several remarks.

1. We conclude from the discussion regarding formula (41)–(49) that the fundamental principle of extremal action remains valid (is preserved) in the case of nonlinear holonomic constraints too.

2. We may try to go over from the Hamiltonian $\mathcal{H}(\wp, q, t)$, defined by (34), to the corresponding Lagrangian $L(\dot{q}, q, t)$ by means of legendre transformation in order to be able, for example, to express the equations of motion of the system with constraints in the form of the exact Euler's equation

$$\frac{d}{dt}\left(\frac{\partial L}{\partial \dot{q}}\right) - \frac{\partial L}{\partial q} = 0. \tag{50}$$

However, it has to be borne in mind that such a Lagrangian (for fixed q and t) will be defined for only those $\dot{q}$ that satisfy the condition (15). This is clear from Eq. (21) which is the "solving" equation for the Legendre transformation. As demonstrated above, this equation yields, for fixed q and t, only those q which automatically satisfy (15) for an arbitrary $\wp$. Therefore for a $\dot{q}$ not satisfying (15) the corresponding $\wp$ cannot be found. Thus the equation cannot be solved for $\wp$ with an arbitrary $\dot{q}$. This equation is the one that leads to condition (15).

3. It is seen from formula (39) that if the second member (the one in the parenthesis) in the right side of this equation vanishes without obeying (16), we would obtain exact Hamiltonian equations for the nonholonomic system with the Hamiltonian (20) or (34).

However, this member of (39) does not, in general, vanish even for linear nonholonomic constraints, what to talk of nonlinear nonholonomic constraints.

It is not plain a priori whether of not there exist *exact* hamiltonian equations (and hence *exact* Hamiltonians) for nonholonomic mechanical systems. It would be interesting to have an answer to this question.

32

CDS Subject to Phase Constraints

The method of phase portrait enables us to take into view constructively the presence of phase constraints, that is, constraints imposed on the state of the CDS. Geometrically, the presence of constraints on the state implies that the representative point q of the given CDS cannot lie in an isolated given set G in the state space { q }. As a rule, $G \subset \{ q \}$ is prescribed *a priori* independently of the basic equation of the CDS.

$$\dot{q} = f (q, u), \qquad u \in U (q). \tag{1}$$

When the condition $q \bar{\in} G$ is present, the investigation of (1) is again carried out by means of its phase portrait. Namely, the set G is superposed on the phase portrait of (1) drawn without taking into consideration the condition that $q \bar{\in} G$. Relatively simpler is to investigate a CDS defined on two-dimensional manifolds, and, in particular, on the two-dimensional plane. Nontrivial will be the case of the CDS (1) subject to the condition $q \bar{\in} G$ when the set G lies entirely or partially outside the domain of unconstrained (free) motions of the given CDS (1). Fig. 32.1 depicts a typical case where the set G (marked by dashes) has been superposed on the phase portrait of CDS (1) drawn first without taking into account the condition $q \bar{\in} G$. In order to determine the admissible trajectory (in accordance with hatching of boundaries of the trajectory funnels and not intersecting the prohibited set G), it is necessary and sufficient to determine a bypass from the initial point q_0 to the final point q_1. In Fig. 32.1, such admissible paths are $q_0 \, E \, q_1$ and $q_0 \, F \, q_1$. However, as is clear from the phase portrait, if the representative point of the CDS falls inside the curvilinear triangle ABC (that is, inside the trajectory funnel with vertex A), then it cannot leave the triangle ABC no matter what admissible path it chooses to do so ; as if the representative point has been "arrested" by the domain ABC. In analogy with astrophysics, such a domain ABC could be termed as a "black hole". The representative point can only enter this domain but can never leave it.

On the other hand, the domain BCD has a diametrically opposite property. Namely, the representative point q of the CDS (1) cannot enter this domain but it can leave it. Such a domain BCD can be naturally termed as a "white hole".

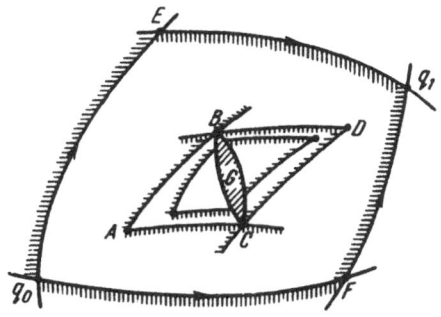

Fig. 32.1

Thus a sailent feature of a CDS subjected to constraints on the state coordinates is the presence of black and white hole type domains. There arises a natural problem of constructing and describing these domains. It is clear that this problem can be easily solved by geometrical (graphical) methods for the case of two-dimensional manifolds.

In the general case, when G does not lie in the domain of unconstrained motions, to determine black and white hole type domains one has to solve, respectively, the Cauchy problem for partial differential equations : $H (\partial z / \partial q, q) = 0$, $H (- \partial z / \partial q, q) = 0$; the initial sets are definite subsets of the prohibited set G.

33

Singular Sets of Two-Dimensional CDSs

For a two-dimensional CDS the singular set will be, as a rule, a curve on a two-dimensional manifold (in particular, on the plane) in the state space { q } of the given CDS. It is clear from the general discussion that such a curve is characterized by a number of following equivalent conditions.

1. $P(p, q, u) = p f(q, u)$ is independent of $u \in U$.

2. $\frac{\partial p}{\partial u}(p, q, u) = 0$ identically in $u \in U$.

3. The equation

$$H(p, q) + H(-p, q) = 0$$

is satisfied.

4. On account of $H(p, q) = 0$, Condition 3 yields the following equivalent system of equations :

$$\begin{cases} H(p, q) = 0 \\ H(-p, q) = 0. \end{cases}$$

5. For a two-dimensional system another condition, equivalent to these mentioned above, can be added. Namely, all the vectors $\dot{q} = f(q, u)$, where u runs through U, are collinear.

For the general two-dimensional system, the most "direct", if one may so express, for the singularity of a curve is Condition 4. This condition simply yields a definite system of equations which can be used, the system is homogeneous, for eliminating the vector $p = (p_1, p_2)$ and obtaining a direct relation between the coordinates q^1 and q^2 of the point q lying on the singular curve.

Indeed, from (4) we have the system

$$\begin{cases} H(p_1, p_2, q^1, q^2) = 0, \\ H(-p_1, -p_2, q^1, q^2) = 0, \end{cases} \tag{A}$$

Since $H(p, q)$ is positive and homogeneous in p, we can impose, with no loss of generality, an additional condition on p. For example, take $|p| = 1$. Eliminating p_1 and p_2 from this system, we obtain the desired equation of the singular curve

$$S(q^1, q^2) = 0. \tag{B}$$

The singular curve is obtained most simply when the CDS is of the form

$$\begin{cases} \dot{q} = f_0(q) + u\,f_1(q), \\ u \in U(q) \subset R^1, \qquad q \in R^2, \end{cases} \tag{C}$$

Applying one of the Conditions 1 - 5, we easily find that the singular curve is determined by the condition of vanishing of the determinant $s(q)$:

$$S(q) \equiv |f_0(q) \quad f_1(q)| = 0. \tag{D}$$

In particular, for the linear system $\dot{q} = Aq + bu$ we have $s(q) \equiv |Aq \quad b| = 0$, and for the bilinear system $\dot{q} = Aq + uBq$ we have $s(q) \equiv |Aq \quad Bq| = 0$.

Phase Portrait of CDS on Two-Dimensional Manifolds

If the representative point of a CDS moves on a two-dimensional manifold M^2, for example, on a two-dimensional plane, two-dimensional sphere or torus etc, the construction of the phase portrait is especially simple.

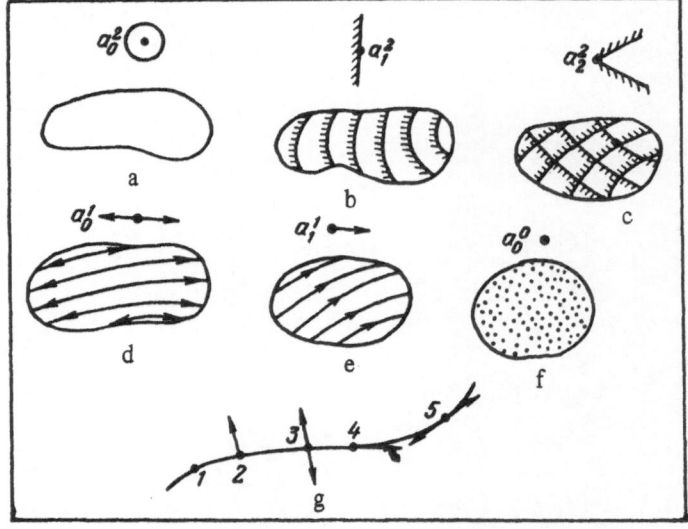

Fig. 34.1

For a two-dimensional CDS there are only six types of cones K (q). Figs. 34.1 a - f depict all of them. These six types of cones correspond to the first three rows of Table 8.1.

Fig. 34.1 g depicts an isolated singular curve with all possible types of cones on it and five cases of their location relative to the singular curve. The motion of such a CDS is represented by a system of two scalar equations

$$\begin{cases} \dot{q}^1 = f^1 (q^1, q^2, u), \\ \dot{q}^2 = f^2 (q^1, q^2, u), \end{cases} \tag{1}$$

where the values of the control u lie in the given set U which, in general, depends on the point $q = (q^1, q^2)$, that is, $U = U (q)$. The system (1) is equivalent to the differential inclusion

$$\dot{q} \in f (q, U) \equiv \Phi (q), \tag{2}$$

where $\Phi (q)$, for each fixed $q \in M^2$, is the convex hull of the mapping $U \longrightarrow T_q M^2$ (when u ranges over the entire set $U (q)$). Complete information regarding CDS (1), (2) is possessed by the Hamiltonian $H (p, q)$ of this system or, what is same, by the support function

$$H (p, q) \equiv \sup (\max) [p_1 f^1 (q^1, q^2, u) + p_2 f^2 (q^1, q^2, u)] =$$
$$\quad u \in U (q)$$

$$= \sup (\max) \quad p\dot{q} \tag{A}$$
$$\dot{q} \in \Phi (q)$$

of the set $\Phi (q)$.

In order to construct the phase portrait of CDS (1), it is necessary to identify its elements, including the singular curve.

The order in which the elements of the phase portrait are identified is, naturally, immaterial since in the final analysis all information is depicted in one figure, the phase portrait. For a particular CDS some of the elements of its phase portrait may be absent. For example, the domain of unconstrained (free) trajectories may not exist or, say, the points of absolute equilibrium may not be present and so on.

It should be remarked that for certain types of CDSs the identification of elements of its phase portriat can be carried out directly with the aid of the right hand side of its differential equation or inclusion.

Let us examine in some detail some of the elements of the phase protrait.

The set of the points $q \in M^2$ of absolute equilibrium (rest) of CDS (1) is determined by the condition

$$\Phi (q) = \{ 0 \} \quad \text{or} \quad | f (q, u) | = 0 \tag{3}$$

independently of $u \in U (q)$. This condition defining the set of points of absolute equilibrium is applicable to CDS of any dimension n. It is plain that the set of points of absolute equilibrium of CDS will be its invariant set in $\{ q \}$.

It proves useful sometimes to identify the set of points $q \in M^2$ for which there exist $u \in U (q)$ satisfying

$$| f (q, u) | = 0 \tag{4}$$

This set will be known as the *set of relative equilibrium* (rest) of CDS (1). The points of relative equilibrium may differ from those of absolute equilibrium. The definition of relative equilibrium is also applicable to CDS of any dimension n. It is obvious that the set of relative equilibrium contians the set of absolute equilibrium of the CDS.

The domain D of free trajectories of CDS (1) (Sec. 9) is defined with the aid of H (p, q) by the inequality

$$\min_{|p| = 1} \quad H (p, q) > 0. \tag{5}$$

This condition is equivalent to the condition that the point $\dot{O} (q) \in T_q M^2$ lies inside $\Phi (q)$ for any $q \in D$ or, what is same, to the condition that the cone $K (q)$ coincides with $T_q M^2$. For two-dimensional systems, this geometrical condition is simpler than its analytical counterpart (5) since for the former computation using formula (5) is not required.

The singular curve S, and, in particular, the curve of reverse hatching is determined by the system of equation (Sec. 17)

$$H (p, q) = 0, \quad H (- p, q) = 0. \tag{6}$$

We eliminate p from (6), bearing in mind that the function H (p, q) is homogeneous in p, and obtain the equation

$$S (q) = 0 \tag{7}$$

of S.

It is essential to underline that (7) indeed defines a proper *curve* provided (7) is satisfied *non-identically* in a domain $G \subset \{ q \}$. For the linear system

$$\dot{q} = A q + bu, \quad u \in U (q), \tag{8}$$

where $q = (q^1, q^2)$ and u is a scalar, the singular curve, and, in particular, the curve of reverse hatching, is a straight line

$$S (q) \equiv | A q \quad b | = 0. \tag{9}$$

For the bilinear system

$$\dot{q} = A q + u B q, \quad u \in U (q), \tag{10}$$

where again $q = (q^1, q^2)$ and u is a scalar, the singular curve (the curve of reverse hatching, in particular) is given by the equation

$$S (q) \equiv | Aq \quad bq | = 0. \tag{11}$$

Since Eq. (11) in q is homogeneous, it is easily seen that Eq. (11) represents either a pair of straight lines intersecting at $\dot{O}$ or a unique point $\dot{O}$.

For a nonlinear system of the type

$$\dot{q} = f_0(q) + u\, f_1(q), u \in U(q), \tag{12}$$

where $q = (q^1, q^2)$ and u is a scalar, the equation of the singular curve, and, in particular, of the hatching reversal curve, is

$$S(q) \equiv |\, f_0(q) \quad f_1(q)\,| = 0. \tag{13}$$

In Eqs. (9), (11) and (13), vertical bars denote second-order determinants.

At points $q \in M^2$, not lying in the set of absolute equilibrium nor in the domain of free motions D, the cone K (q) is enclosed by two generators, the left generator to which there corresponds a control u_1 (q) and the right to which there corresponds a control u_2 (q). To these generators and controls u_1 (q) and u_2 (q), there correspond left and right branches of the boundaries of trajectory funnel originating from the given point q. The left branch is defined by an equation of the form

$$\dot{q} = f(q, u_1(q)). \tag{8 a}$$

and the right branch by

$$\dot{q} = f(q, u_2(q)). \tag{9 a}$$

If the cone K (q) is not closed, that is, if either the left or the right ray (or both) does not belong to K (q), then two nonzero vectors φ_1 (q) and φ_2 (q), collinear with the generators, are chosen. The system of equations for the boundaries of the trajectory funnels then becomes

$$\dot{q} = f_1(q) \tag{10 a}$$

for the left branch and

$$\dot{q} = f_2(q) \tag{11 a}$$

for the right branch.

We perform hatching on the left branch, described by Eq. (8a), of the boundary of the trajectory funnel from the left side as we move along with the passage of positive time. Similarly, hatching is performed on the right branch from the right side.

As to invariant curves, they are of two kinds : "isolated invariant curves" given by Eq. (7) and "non-isolated invariant curves" which fill up (divide into layers) the domain of { q }.

The *isolated invariant curve* is a subset of the singular curve (7), and for it Cases 4, 5, depicted in Fig. 34.1 g, are necessarily satisfied. That is, K (q) is of type a_1^1 or a_0^1 and *touches* the singular curve. The question whether or not the representative point $\overset{1}{a_0}$ or $\overset{1}{a_1}$ of the CDS can leave a highter order singular curve is, apparently, much easier to deal with directly on the basis of the equation of the CDS.

The domains filled up by invariant curves (Fig. 34.1 d.e) are identified as those domains of the state space { q } for which (7) is satisfied identically, and the cone K (q) is of type $\overset{1}{a_0}$ or $\overset{1}{a_1}$.

The aggregate of all the abovementioned objects on M^2 constitutes the phase portrait of the given CDS.

The main significance of the phase portrait lies in that it enables us to draw all possible admissible trajectories of the given CDS constituted by the action of admissible controls $u \in U (q)$. A trajectory on the phase portrait will be admissible if it "pierces" the boundary of the trajectory funnel in accordance with hatching, that is, from the side where hatching is performed. If the cone K (q) of admissible velocities is closed along the boundary of the funnel, that is, if two of its generators also provide admissible velocity directions, then the motion strictly along the boundaries of the funnel is feasible. That is, the boundaries of the funnels themselves constitute admissible trajectories.

We thus see that we are, in reality, concerned with the investigation of two families of trajectories, the left and right, determined, respectively, by the Eqs. (8a), (9a) or (10a), (11a). In principle, Eqs. (8a), (9a) or (10a), (11a) can be prescribed independently of each other and of the inclusion (2). Any one of them can be called the equation of the left boundary and the other that of right boundary. Thus we are led to the investigation of two independent families of trajectories, superposed on each other, of differential equations

$$\dot{q} = f_1(q), \qquad \dot{q} = f_2(q). \tag{12 a}$$

In investigating two phase portraits, superposed on each other, of differential Eqs. (12a) there arise, apart from the familiar notations of singular points of these equations, a new notion of "singular sets"; these sets reflect the properties of the CDS mentioned in Sec. 33.

Examples on Construction of Phase Portrait of Two-Dimensional CDSs

In order to illustrate the results of the preceding section, let us consider example on constructing phase portraits of CDSs when the state space (the phase space) is the usual two-dimensional plane.

1. For a linear CDS governed by the equations

$$\dot{q}^1 = q^2, \quad \dot{q}^2 = u, \quad |u| \leq m, \quad m > 0, \tag{1}$$

on the plane (q^1, q^2) there are no points of absolute equilibrium, and the points of relative equilibrium are furnished, for $u = 0$, by the points of the q^1-axis. Also absent is the domain of unconstrained free motions. The hatching reversal curve is given here by the equation

$$s \equiv \begin{vmatrix} q^2 & 0 \\ 0 & 1 \end{vmatrix} = q^2 = 0. \tag{2}$$

and is, hence, the q^1-axis. The set of relative equilibrium also coincides with the q^1-axis.

The boundaries (right and left) of trajectory funnels are described by the equations

$$\dot{q}^1 = q^2, \dot{q}^2 = \pm m, \tag{3}$$

where the plus sign in the second equation corresponds to one boundary and the negative sign to the other boundary. Dividing the first equation in (3) by the second, we obtain a single equation.

$$\frac{dq^1}{dq^2} = \pm \frac{q^2}{m}. \tag{4}$$

This equation furnishes directly the boundaries of trajectory funnels on the phase plane (q^1, q^2). Integration of (4) provides a family of parabolas

$$q^1 = \pm \frac{1}{2m}(q^2)^2 + c, \qquad |c| < \infty. \tag{5}$$

on the plane (q^1, q^2). This family of boundaries of trajectory funnels of CDS (1) together with hatching is depicted in Fig. 35.1. The boundary of an individual funnel with vertex A is identified by a thick curve (its right branch is continued only up to the point where it intersects the hatching reversal curve, that is, up to the q^1-axis).

The phase portrait of this system clearly shows that CDS (1) is completely controllable. That is, any two given points q_0 and q_1 on the plane (q^1, q^2) can be joined by an admissible trajectory along which the representative point moves from q_0 to q_1 in a finite time.

2. For the bilinear CDS governed by the equation

$$\dot{q}^1 = q^2 + u q^1, \quad \dot{q}^2 = q^1 + u q^2, \quad |u| \le m, \quad m > 0, \tag{6}$$

on the plane (q^1, q^2) the set of absolute equilibrium is the singleton 0. Observe that

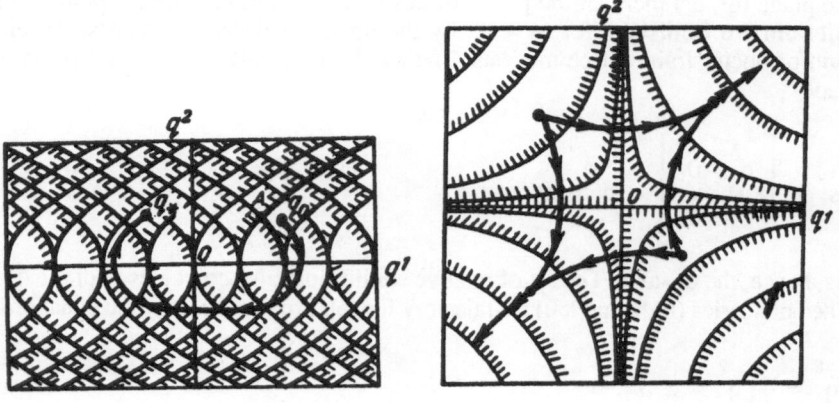

Fig. 35.1   Fig. 35.2

for any bilinear system $\dot{q} = A q + u B q$ the point $q = 0$ is, in general, a point of absolute equilibrium. It can be easily seen that CDS (6) has no point of relative equilibrium distinct from the points of absolute equilibrium. This is because

$$| f(q, u)|^2 \equiv |Aq + uBq| = (q^2 + 4q^1)^2 + (q^1 - uq^2)^2 = O \tag{7}$$

if $q^1 = 0$ and $q^2 = 0$, and this point coincides with the point of absolute equilibrium. If $q \neq O$, there is no $u = u(q)$ so that (7) is satisfied.

The equation of the hatching reversal curve for CDS (6) is of the form

$$s(q) \equiv |AqBq| = \begin{vmatrix} q^2 & q^1 \\ q^1 & -q^2 \end{vmatrix} = -(q^2)^2 - (q^1)^2 = 0. \tag{8}$$

This shows that the hatching reversal curve is absent. It degenerates into the same point O which constitutes the set of absolute equilibrium.

Proceeding as in the previous example, we let $u = \pm m$ and divide the first equation of (6) by the second. This yields the equation

$$\frac{dq^1}{dq^2} = \frac{q^2 \pm m q^1}{q^1 \mp m q^2} \tag{9}$$

for the boundaries of trajectory funnels of CDS (6). We integrate Eq. (9) in the limiting case $m \longrightarrow \infty$ and obtain

$$\frac{dq^1}{dq^2} = -\frac{q^1}{q^2}. \tag{10}$$

Integration of (10) shows that the boundaries of trajectory funnels of CDS (6), as $m \longrightarrow \infty$, are the hyperbolas

$$q^1 q^2 = c, \qquad |c| < \infty. \tag{11}$$

Fig. 35.2 depicts the family (11) together with hatching. The phase portrait of CDS (6) in Fig. 35.2 shows that the given CDS is uncontrollable. It is clear that, for example, there is no admissible trajectory that would join the initial point q_0 located in the first quadrant of the plane (q^1, q^2) with the final point q_1 located in the second quadrant.

3. Consider the CDS which differs from (6) only in the sign of q^1 in the second equation. That is, the CDS is now governed by the equations

$$\dot{q}^1 = q^2 + u q^1, \qquad \dot{q}^2 = -q^1 - u q^2. \tag{12}$$

Let the scalar control u be unbounded. As in Example 2, the point $\dot{O}$ is the only point of absolute equilibrium. And points of relative equilibrium coincide with the points of the hatching reversal curve.

The equation of the hatching reversal curve is of the form

$$s(q) \equiv |AqBq| = \begin{vmatrix} q^2 & q^1 \\ -q^1 & -q^2 \end{vmatrix} = -(q^2)^2 + (q^1)^2 = 0. \tag{13}$$

This equation shows that hatching reversal curves are two straight lines intersecting at the origin and bisecting the coordinate angles of the plane (q^1, q^2).

As in the preceding example, the bounaries of trajectory funnels are the family of hyperbolas (11). The phase portrait of this system is depicted in Fig. 35.3. Unlike CDS (6), the given CDS (12) is completely controllable (in a domain that *excludes the origin*).

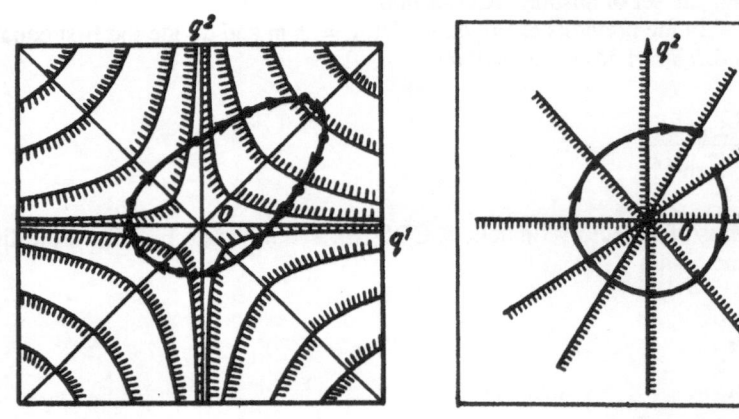

Fig. 35.3   Fig. 35.4

One can explain the controllability of CDS (12) by pointing out that (12) has, in contrast to (6), hatching reversal curves (13) that do not degenerate into a single point. However, as the next example shows, the presence of nondegenerate hatching reversal curves is not at all necessary for the CDS to be completely controllable. We shall discover in Example 4 that the CDS is completely controllable although the hatching reversal curve degenerates into a point, the origin.

It should be remarked that a bilinear system for which O is the point of absolute equilibrium can naturally be termed *completely controllable* although it is obvious that the representative point cannot leave O nor can it enter O in finite time under the action of an admissible control. It is natural therefore to consider controllability of such systems on the manifold (q^1, q^2) without the point q = O (the plane is "punctured" at the origin).

4. Let the bilinear CDS be governed by the equations

$$\dot{q}^1 = q^2 + u q^1, \qquad \dot{q}^2 = -q^1 + u q^2, \tag{14}$$

where u is an unbounded control. Arguing in the same way as in Examples 2, 3, we can easily establish that the phase portrait of the given CDS is of the form depicted in Fig. 35.4. The boundaries of trajectory funnels here are straight lines $q^1 = c q^2, |c| \leq \infty$, but the hatching reversal curve degenerates into the point $\dot{O}$. Notwithstanding such a degeneration of the hatching reversal curve, the CDS is completely controllable in the plane (q^1, q^2) without O.

5. We now consider on the plane the nonlinear CDS governed by the equations

$$\dot{q}^1 = u^1 q^1 q^2, \qquad \dot{q}^2 = (u^1)^2 q^1. \tag{15}$$

The set of absolute equilibrium here is furnished by the line $q^1 = 0$, the q^2-axis. The set of relative equilibrium is the remainder plane (q^1, q^2) with $u = 0$.

The convex set of admissible velocities f (q, U) of this CDS is shown by dashes in Fig. 35.5. It has been drawn for the points (q^1, q^2) with $q^1 \neq 0$, $q^2 \neq 0$ (Fig. 35.5 depicts, in particular, the case $q^1 > 0$ and $q^2 > 0$). The curvilinear boundary of this set is a parabola with vertex $\dot{O}$ (q^1, q^2) showing that the point 0 always lies on the boundary of the domain f (q, U). Consequently, the domain of unconstrained (free) motions is here absent. The closure of the cone K (q) for the points (q^1, q^2), with $q^1 \neq 0$ and $q^2 \neq 0$, is the half-plane bounded by a line parallel to the q^1-axis. This line is tangent to f (q, U) at the point $\dot{O}$ (q^1, q^2). The aggregate of lines $q^2 = c$, | c | $< \infty$, gives the family of boundaries of trajectory funnels. The phase portrait of this system is depicted in Fig. 35.6.

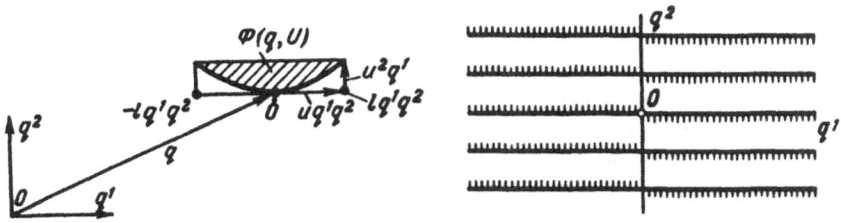

Fig. 35.5   Fig. 35.6

It has to be underlined that in accordance with the form of the convex set f (q, U) for points q ≠ 0 on the q^1-axis the cone K (q) degenerates into a ray. This ray is directed vertically upward in the right half-plane and vertically downward in the left half-plane. This implies that the admissible trajectory necessarily intersects the q^1-axis at a right angle.

Furthermore, the structure of f (q, U) shows that the representative point cannot move strictly along boundaries of trajectory funnels (that is, along the q^1-axis) as it is clear from Fig. 35.5 that the velocity $\dot{q}$ in this direction is zero. However, at any angle (even very small) to the boundary motion is feasible in the admissible direction.

Finally, the phase portrait drawn in Fig. 35.6 shows that although the q^2-axis is the hatching reversal curve yet the representative point cannot intersect it in a finite time. This is due to the fact that, as mentioned earlier, q^2 -axis is the set of absolute equilibrium, that is, an invariant manifold.

The present example thus shows that there exists trajectories for the given CDS which lead, say, from the right half-plane ($q^1 > 0$) to the left-plane ($q^1 < 0$) and satisfy the principle of inclusion in the state space (Sec. 13). Yet, such a trajectory cannot be admissible because the condition of finite time, condition (13.3), is not satisfied.

Thus CDS (15) is uncontrollable. All the admissible motions of its representative point can be described as follows. In the right half-plane $q^1 > 0$ it can move only from bottom upwards and sideways ; movements from top to bottom and in horizontal direction are ruled out. The representative point cannot reach the q^2 -axis and it can intersect the q^1 -axis only at a right angle. In the left half-plane $q^1 < 0$ the picture is similar. The representative point can move only from top to bottom and sideways but cannot move horizontally and from bottom upwards. Here too the q^1 -axis can be intersected only at a right angle and the q^2 - axis is inaccessible for the representative point.

6. Consider the bilinear CDS whose equations of motion are

$$\dot{q}^1 = -a\,q^2, \quad \dot{q}^2 = a\,q^1 + u\,q^3, \quad \dot{q}^3 = -u\,q^2, \quad |u| \leq m, \tag{16}$$

where the quantities $a > 0$, $m > 0$ and u is a scalar. Although this system formally consists of three equations, yet, in view of the presence of the first integral

$$F(q) \equiv (q^1)^2 + (q^2)^2 + (q^3)^2 = r^2, \tag{17}$$

where r is an arbitrary constant, the representative point of CDS (16) will move only on a two-dimensional sphere. Integral (17) can be derived in a simple manner by multiplying the first equation in (16) by q^1, the second by q^2 and the third by q^3 and then adding the three equations. This results in zero on the right hand side and a complete differential of $F(q)$ in (17) on the left hand side, and this leads to (17). Eqs. (16) have a direct physical significance. They describe the motion of spin in a magnetic field, and are known as the *Bloch's equations* [30].

Thus, by virtue of (17), this systems has invariant manifolds and the representative point q of CDS (16) moves on the two-domensional sphere with centre O, given by (17). The constant r plays the role of the radius of this sphere and is determined by the initial point q_0.

When the invariant manifold is a sphere with centre O, it is convenient to move over to spherical coordinates using the formulas

$$q^1 = r \sin\theta \, \cos\varphi, \quad q^2 = r \sin\theta \, \sin\varphi, \quad q^3 = r \cos\theta, \tag{18}$$
$$0 \leq \varphi < 2\pi, \quad 0 \leq \theta \leq \pi.$$

Substituting (18) into (16), we find that the original CDS is governed by two equations

$$\dot{\varphi} = a + u \operatorname{ctg}\theta, \quad \dot{\theta} = u \sin\varphi. \tag{19}$$

It is clear that for the given CDS the points of absolute rest are absent. Also obvious

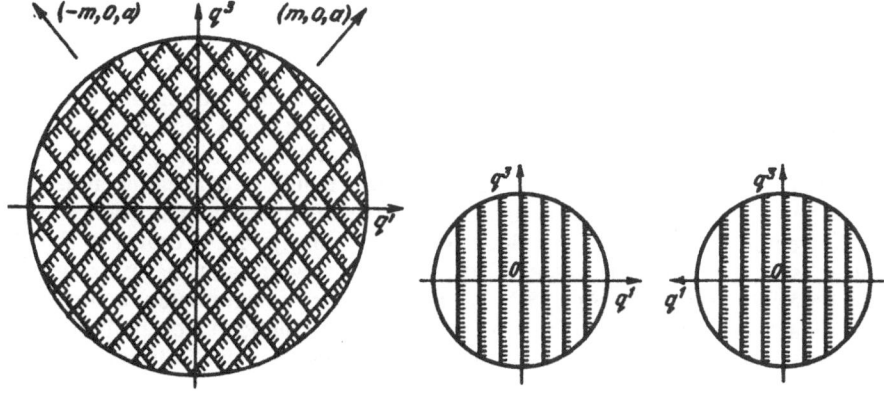

Fig. 35.7   Fig. 35.8

is the absence of the domain of free trajectories. The hatching reversal curve on the sphere is determined by the condition

$$s(\varphi, \theta) = \begin{vmatrix} a & ctg\,\theta \\ 0 & \sin\varphi \end{vmatrix} = a \sin \varphi = 0. \qquad (A)$$

Since a > 0, the last condition yields $\sin \varphi = 0$. Hence $\varphi = 0$ or $\varphi = \pi$, in any case, determines only one curve, namely, a great circle of the sphere lying in the plane (q^1, q^3).

The boundaries of trajectory funnels on the invariant sphere are given by Eqs. (19) with $u = \pm m$:

$$\dot\varphi = a \pm m\,ctg\,\theta, \qquad \dot\theta = \pm\,m \sin \varphi. \qquad (20)$$

Dividing the first equation of (20) by the second, we obtain the differential equation

$$\frac{d\varphi}{d\theta} = \frac{a \pm m\,ctg\,\theta}{\pm\,m \sin \varphi} \qquad (21)$$

for the family of boundaries of trajectory funnels. It is an equation with variables separable and its integration yields finally the equations

$$\pm \cos \varphi = \beta\,\theta + \ln|\sin \theta| + c, \qquad (22)$$

for the family of boundaries of trajectory funnels in the coordinates (φ, θ), here r = const., the constant parameter $\beta = a/m$ amd c is an arbitrary constant.

If the phase portrait of the given CDS is drawn on the plane (φ, θ), or, more precisely, on the segment of the right circular cylinder $0 \le \varphi < 2\pi,\ 0 \le \theta \le \pi$, with perimeter of the base equal to 2n and height π, then boundaries of trajectory funnels of system (20) will be curves described by transcendental equations (22).

However, it turns out much simpler to describe the boundaries of trajectory funnels directly on the invariant sphere itself if one notes that the motion of the representative point under the action of the control $u = \pm m$ is a clockwise motion in the plane perpendicular to the vector $(\pm m, 0, a)$ (viewed from the endpoint of this vector). Thus the boundaries of trajectory funnels of CDS (16) are a family of concentric circles situated on the invariant sphere which are obtained by intersecting the sphere by planes perpendicular to the vector $(\pm m, 0, a)$. The phase portrait of this system is drawn in Fig. 35.7 ; the sphere is depicted from the side of the q^1 -axis. The view from the q^2 -axis side is similar. It can be easily seen from Fig. 35.7 that the CDS (16) is completely controllable on the invariant sphere. When $m \longrightarrow \infty$, that is, when u is unbounded, the boundaries of the funnels "straighten" and turn into a single family of circles lying in the plane perpendicular to the q^1 -axis (Fig. 35.8).

7. We now consider a system

$$\dot{q}^1 = q^2 + u^1, \quad \dot{q}^2 = -q^1 + u^2, \qquad |u^{-1}| \le m_1, |u^2| \le m_2, \tag{A}$$

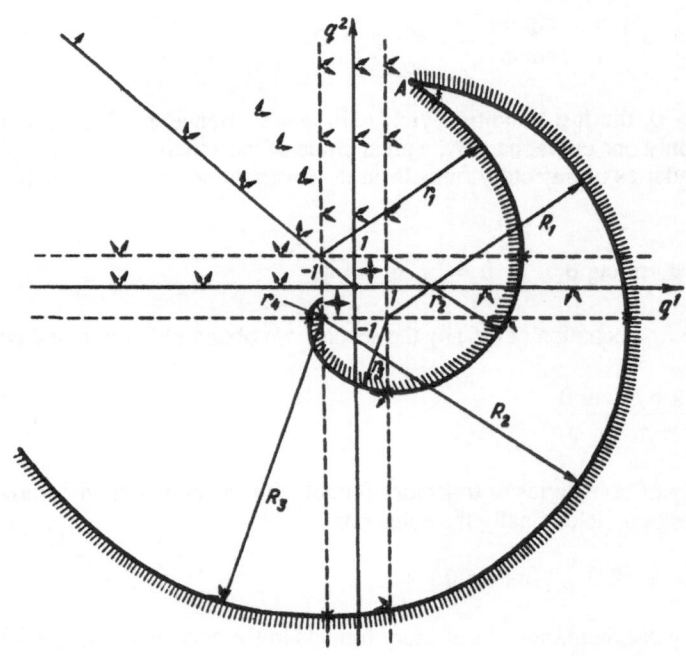

Fig. 35.9

with two controls in u^1 and u^2. Here the set of absolute equilibrium is absent. But the domain D of free trajectories is present and it coincides with the set of relative equilibrium. It is parallelogram with the origin of the (q^1, q^2) - plane as centre and sides of lengths 2 m_1 and 2 m_2 oriented parallel to the respective coordinate axis. This parallelogram (domain D) generates a cross (fig. 35.9) limiting the boundary curves of the funnel. The boundaries of trajectory funnels are arcs of circles of suitable radii depicted in Fig. 35.9. There trajectories in reality represent circles because the given CDS is a controllable harmonic oscillator. The figure depicts a trajectory funnel with vertex A. Funnels with other vertices can be similarly drawn.

Phase Portrait of Two-Level Quantum-Mechanical CDS*

A number of quantum-mechanical objects can be represented as a two-level system governed by the equation [30]

$$i\, h\, \dot{q} = \mathcal{H} q + u(t)\mathcal{H}_1 q. \tag{1}$$

Here q denotes a two-dimensional complex vectors, $\mathcal{H}$, $\mathcal{H}_1$ denote Hermitian matrices, u(t) a real piecewise continuous number function and n denotes the Planck constant. For this system the normalizing condition $|q| = 1$ holds. It corresponds to the interpretation of the complex quantities q^1 and q^2 as probability amplitudes of the object attaining ground states 11 > and 12>, respectively. An essential characteristic of the state of the object is the amplitude *ratio* $q^1(t)/q^2(t)$. The amplitude ratio completely characterizes the ratio of probabilities of the state of the quantum system in two ground states 11 > and 12 > and their phase difference.

Let us formulate the corresponding problem of finite control [23, 24]. Find a piecewise continuous control u(t) such that for some finite $T \geq 0$ and prescribed complex quantities k_0 and k_1 the solution $q = q(t)$ of system (1) satisfies the conditions

$$\frac{q^1(0)}{q^2(0)} = k_0, \tag{2}$$

$$\frac{q^1(T)}{q^2(T)} = k_1, \tag{3}$$

---

* The results of the present section are due to A.V. Babichev.

For the initial point of such a solution any point ($\overset{1}{q_0}$ $\overset{2}{q_0}$) satisfying $\overset{1}{q_0}/\overset{2}{q_0} = k_0$ can be chosen. By a unitary transformation with the matrix $R = \| \overset{i}{r_j} \|$ the system (1) can be transformed to a system

$$i h \dot{\tilde{q}} = \tilde{\mathcal{H}} \tilde{q} + u(t) \tilde{\mathcal{H}}_1 \tilde{q},$$

(4)

with a real diagonal matrix $\mathcal{H}_1$ in the term containing the control. Here $\tilde{\mathcal{H}} = R \mathcal{H} R^{-1}$ and $\tilde{\mathcal{H}}_1 = R \tilde{\mathcal{H}}_1 R^{-1}$ is a real diagonal matrix and $\tilde{q} = Rq$. Conditions (2), (3) then assume the form

$$\frac{\tilde{q}^1(0)}{\tilde{q}^2(0)} = \tilde{k}_0, \qquad \tilde{k}_0 = \frac{\overset{1}{r_1} k_0 + \overset{1}{r_2}}{\overset{2}{r_1} k_0 + \overset{2}{r_2}},$$

(5)

$$\frac{\tilde{q}^1(T)}{\tilde{q}^2(T)} = \tilde{k}_1, \qquad \tilde{k}_1 = \frac{\overset{1}{r_1} k_1 + \overset{1}{r_2}}{\overset{2}{r_1} k_1 + \overset{2}{r_2}},$$

(6)

We introduce the variables ρ, φ for the ratio of the complex amplitudes by the formula

$$\frac{\overset{1}{q}(t)}{\overset{2}{q}(t)} = \rho(t) e^{i \varphi(t)}$$

(7)

Then Eqs. (4)-(6) yield the following equations in ρ, φ and φ:

$$\dot{\rho} = (\rho^2 + 1)(a \cos \varphi + b \sin \varphi),$$

(8)

$$\dot{\varphi} = \left(\rho - \frac{1}{\rho} \right)(b \cos \varphi - a \sin \varphi) + d + g u(t),$$

$$\rho(0) = \rho_0 = |\tilde{k}_0|, \varphi(0) = \varphi_0 = \arg \tilde{k}_0,$$

(9)

$$\rho(T) = \rho_1 = |k_1|, \varphi(T) = \varphi_1 = \arg \tilde{k}_1$$

(10)

where a, b, d, g are real quantities. It can be shown that

$$a \neq 0, b \neq 0, g \neq 0$$

(11)

provided the matrices $\mathcal{H}$ and $\mathcal{H}_1$ of the initial system (1) satisfy the condition

$$[\mathcal{H}, \mathcal{H}_1] \equiv \mathcal{H} \mathcal{H}_1 - \mathcal{H}_1 \mathcal{H} \neq 0.$$

(12)

Let us draw the phase portrait of CDS (8), (9). Let $\mid u\,(\,t\,)\mid \,\le\, 1$. We shall draw the boundary curves, given by the equation $z\,(\,\rho,\,\varphi\,)\,=\,0$, of the trajectory funnel on the manifold $\{\,\rho,\,\varphi\,\}$. Following the method for boundaries of trajectory funnels, we introduce the Pontryagin's function

$$P\,(\,p_1, p_2, \rho, \varphi, u\,)\,\equiv\,p_1\,(\,\rho^2\,+\,1\,)\,(a\cos\varphi\,+\,b\sin\varphi\,)\,+$$

$$+\,p_2\left[\left(\rho-\frac{1}{\rho}\right)(b\cos\varphi\,+\,a\sin\varphi)\,+\,d\right]\,+\,ugp_2.\tag{A}$$

for the CDS (8), (9). For this function

$$\begin{array}{c}\arg\max \\ \mid u\mid\,\le\,1\end{array} P\,(\,p_1, p_2, \rho, \varphi, u\,)\,=\,1\,\text{sign}\,(\,gp_2\,).\tag{B}$$

The differential equation in $z\,(\,\rho,\,\varphi\,)$ then becomes

$$\frac{\partial z}{\partial \rho}\,(\rho^2\,+\,1)\,(\,a\cos\varphi\,+\,b\sin\varphi\,)\,+\,\frac{\partial z}{\partial \varphi}\left(\rho-\frac{1}{\rho}\right)$$
$$(\,b\cos\varphi\,-\,a\sin\varphi\,)\,+\,d\,+\,\lg\,\text{sign}\left(\frac{\partial z}{\partial \varphi}\,g\right)\bigg]\,=\,0\tag{13}$$

or

$$\frac{\partial z}{\partial \rho}\frac{1}{l}(\rho^2\,+\,1\,)\,(\,a\cos\varphi\,+\,b\sin\varphi\,)\,+$$

$$+\,\frac{\partial z}{\partial \varphi}\left[\frac{1}{l}\left(\rho-\frac{1}{\rho}\right)(\,b\cos\varphi\,-\,a\sin\varphi\,+\,d\right]\,+\,\left|\frac{\partial z}{\partial \varphi}\,g\right|\,=\,0.\tag{14}$$

Passing to limit in (14) as $l\,\longrightarrow\,\infty$, we obtain the equation

$$\frac{\partial z}{\partial \varphi}\,g\,=\,0.\tag{15}$$

In view of (12), $g\,\ne\,0$, and Eq. (15) has the unique solution $\partial z\,/\,\partial\varphi\,=\,0$.

Treating ρ and φ as polar co-ordinates, we find from (15) that the boundaries of trajectory funnels $z\,(\,\rho,\,\varphi\,)\,=\,0$ are circles with centre at the origin (Fig. 36.1).

The singular set of the system is determined by the equation

$$\begin{vmatrix}(\rho^2\,+\,1)\,(a\cos\varphi\,+\,b\sin\varphi) & 0 \\ \left(\rho-\frac{1}{\rho}\right)(b\cos\varphi\,-\,a\sin\varphi)\,+\,d & g\end{vmatrix}\,=\,0.\tag{16}$$

By virtue of (12), all solutions of Eq. (16) are of the form

$$\varphi^{k} = k\pi - \text{ar c tg}\left(\frac{a}{b}\right), k = 0, \pm 1, \dots\dots$$ (17)

Hatching on the circle $z(\rho, \varphi) = 0$ can be easily determined by Eq. (8). It is clear that the

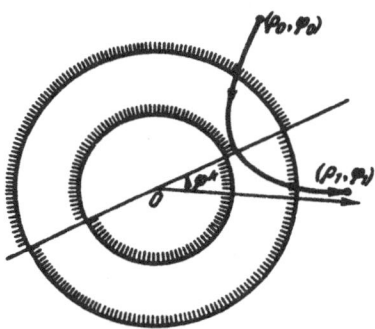

Fig. 36.1

singular set of the system (8), (9) is the manifold of reverse hatching as well.

It should be noted that under conditions (10) = ρ_0, φ_1 can assume the values 0 or ∞. But the corresponding values of φ_0 and φ_1 remain undetermined. The transition between such initial and final points can be performed in a finite time with a bounded control u(t). For instance, such a transition can be accomplished if one moves along a ray directed at an angle $\alpha_m = m\pi + \text{arc t g}(b/a)$, m = 0, ± 1,.... with the control u (t) = cons8t. over [0, T].

Systems of type (1) are often considered in physics. To control u(t) there corresponds, as a rule, the intensity of rapidly oscillating electrical or magnetic fields. Therefore it becomes interesting to examine whether the method of phase portrait can be shortened for special classes of admissible controls.

Example of Decomposable Bilinear CDS in Three-Dimensional Space *

Some bilinear CDSs in the three-dimensional space can be decomposed into a system involving control in equations for two variables and a differential equation for the third variable not containing the control. Under certain conditions, the investigation of controllability of the system of two equations with control enables us to draw conclusion regarding controllability of the original bilinear CDS in the three-dimensional space. For example, by this method we can investigate the controllability of the CDS governed by the equation.

$$\dot{x} = \begin{bmatrix} k & 0 & 0 \\ 0 & k & 0 \\ 0 & 0 & r \end{bmatrix} x + u(t) \begin{bmatrix} 0 & 1 & 1 \\ -1 & 0 & 1 \\ -1 & -1 & 0 \end{bmatrix} x, \quad x \in R^3, |u| \leq \infty, \tag{A}$$

where $u(t)$ is a piecewise continuous control and k and r are real numbers.

It can be shown that this system is controllable in R^3 if $-2 < r/k$ 0. By a change of variables given by $q^1 = x^1/x^3, q^2 = x^2/x^3$, the original system can be decomposed. This results in the following system of two equations

$$\dot{q}^1 = a q^1 + u(t)[(q^1)^2 + q^1 q^2 + q^2 + 1],$$
$$\dot{q}^2 = a q^2 + u(t)[(q^2)^2 + q^1 q^2 + q^1 + 1], \tag{1}$$

where $a = k - r$, with control.

Let us draw the phase portrait of CDS (1). Employing the notation

---

* The results of the present section are due to A.V. Babichev.

$$a \begin{bmatrix} q^1_2 \\ q^2 \end{bmatrix} = f(q), \tag{2}$$

$$\begin{bmatrix} (q^1)^2 + q^1 q^2 + q^2 + 1 \\ (q^2)^2 + q^1 q^2 - q^1 + 1 \end{bmatrix} = g(q), q \in R^2, \tag{3}$$

we can write (1) in the form

$$\dot{q} = f(q) + u g(q). \tag{4}$$

$$\arg \max_{|u| \le 1} p[f(q) + ug(q)] = l \, \text{sign}(p g(q)) \tag{5}$$

for any $p \in R^2$. The function $z(q)$ in the equation of the boundary of trajectory funnel $z(q) = 0, \partial z / \partial q = p$ is then determined (Sec. 12) from the equation

$$\frac{\partial z}{\partial q} f(q) + 1 \left| \frac{\partial z}{\partial q} g(q) \right| = 0. \tag{6}$$

or from

$$\frac{1}{l} \cdot \frac{\partial z}{\partial q} f(q) + \left| \frac{\partial z}{\partial q} g(q) \right| = 0. \tag{7}$$

We shall draw the phase portrait for the unbounded control $u(t)$. Passing to limit, as $l \longrightarrow \infty$, in (7) we obtain the equation

$$\frac{\partial z}{\partial q_1} [(q^1)^2 + q^1 q^2 + q^2 + 1] + \frac{\partial z}{\partial q_2} [(q^2)^2 + q^1 q^2 - q^1 + 1] = 0. \tag{8}$$

$$\frac{\partial z}{\partial q_1} [(q^1 + q^2) q^1 - (-1 - q^2)] + \frac{\partial z}{\partial q_2} [(q^1 + q^2) q^2 - (-1 + q^1)] = 0. \tag{9}$$

for the boundaries of trajectory funnel. This is the *Hesse equation*. A method for solving this equation can be found in [51]. Introducing the homogeneous co-ordinates

$$q^1 = \frac{\xi^1_1}{\xi^0}, \qquad q^2 = \frac{\xi^2}{\xi^0}, \tag{10}$$

$$z(q^1, q^2) = \xi(\xi^0, \xi^1, \xi^2) \tag{11}$$

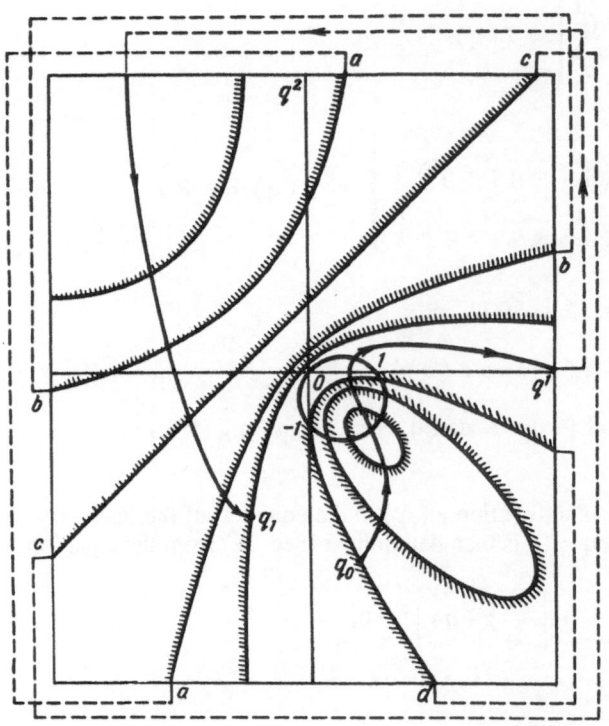

Fig. 37.1

we have

$$\frac{\partial z}{\partial q^1} = \xi^0 \frac{\partial \zeta}{\partial \xi^1}, \qquad \frac{\partial z}{\partial q^2} = \xi^0 \frac{\partial \zeta}{\partial \xi^2}, \qquad q^1 \frac{\partial z}{\partial q^1} + q^2 \frac{\partial z}{\partial q^2} = -\xi_0 \frac{\partial \zeta}{\partial \xi^0} \tag{12}$$

In homogeneous co-ordinates, Eq. (9) becomes

$$(\xi^1 + \xi^2) \frac{\partial \zeta}{\partial \xi^0} - (\xi^0 + \xi^2) \frac{\partial \zeta}{\partial \xi^1} - (\xi^0 + \xi^1) \frac{\partial \zeta}{\partial \xi^2} = 0. \tag{13}$$

The characteristic system for this equation is

$$\xi^0 = \xi^1 + \xi^2;$$

$$\xi^1 = -\xi^0 - \xi^2; \qquad \xi^2 = \xi^1 - \xi^0. \tag{14}$$

which has two independent integrals.

$$F^1 = \xi^0 + \xi^1 - \xi^2 ; \tag{15}$$

$$F^2 = (\xi^0)^2 + (\xi^1)^2 + (\xi^2)^2. \tag{16}$$

We seek an integral F (F^1, F^2) which, on substitution of (10), becomes a function of q^1 and q^2 only. For F we can take the function

$$F = \frac{F^2}{(F^1)^2} + c. \tag{17}$$

On account of (10), (11), the equation z (q) = O of boundaries of trajectory funnels becomes

$$(q^1)^2 (c - 1) + (q^2)^2 (c - 1) - 2c\,q^1q^2 + 2c\,q^1 - 2cq^1 - 2cq^2 + c - 1 = 0. \tag{18}$$

Eq. (18) has following real solutions : (i) a point if $c = \frac{1}{3}$, (ii) an ellipse if $\frac{1}{3} < 0 < \frac{1}{2}$, (iii) a parabola if $c = \frac{1}{2}$, (iv) a hyperbola if and (v) a straight line if $\frac{1}{2} < c < \infty$. When $c < \frac{1}{3}$, Eq. (18) does not have a real solution.

The above mentioned manifolds are boundaries of trajectory funnels for system (1). They have been shown in Fig. 37.1. This figure depicts the phase portrait of CDS governed by Eqs. (1).

Let us establish a correspondence between points at infinity of the phase portrait of (1). This becomes necessary in view of the fact that there are controls u (t) such that the solutions of system (1) go off to infinity in a finite time. The curves determined by Eq. (18) are images of curves in the space (ξ^0, ξ^1, ξ^2) determined by the equations F^1 = const., F^2 = const. under the transformation (10). Since the equations F^1 = const. and F^2 = const. determine closed curves (circles) in the three-dimensional spaces, the aforementioned correspondence between points at infinity of the phase portrait is established, in view of (10), as follows.

1. The points at infinity of the parabola obtained from (18) (c = $\frac{1}{2}$) are identified.

2. Points at infinity lying on different branches of the hyperbola (c > $\frac{1}{2}$) and corresponding to the same asymptote of the hyperbola are identified in pairs.

3. The points at infinity of the line (c = ∞) are identified.

The above correspondence on the phase portrait for the parabola, one of the hyperbolas and one of the lines is shown in the figure by dotted lines beyond the frame of the phase portrait (the identified points have been denoted by identical Latin letters beyond the frame of the phase portrait).

Let us find the singular set for system (1). Points belonging to this set are determined by the equation

$$\det [f (q), g (q)] = 0. \tag{19}$$

Substituting for f (q) and g (q) from (2) and (3), we obtain

$$a [(q^1)^2 + (q^2)^2 q^1 - (q^1)^2 + q^1] - a [(q^1)^2 q^2 + q^1 (q^2)^2 + (q^2)^2 + q^2] = 0, \tag{20}$$

or

$$\left(q^1 - \frac{1}{2} \right)^2 + \left(q^2 - \frac{1}{2} \right)^2 = \frac{1}{2}. \tag{21}$$

Thus the singular set is a circle of radius $1/\sqrt{2}$ with centre $\left(\frac{1}{2}, -\frac{1}{2} \right)$.

Hatching on the boundaries of trajectory funnels can be easily determined in the following manner. For. (8) the characteristic system is of the form

$$\dot{q} = g (q). \tag{22}$$

It is clear that the curves z (q) = 0 are trajectories of this system. Therefore hatching on the boundaries of trajectory funnels z (q) = 0 for system (1) can be determined by superposing the phase portrait of (22) (that is, drawing of curves (18)) and the phase portrait of the system

$$\dot{q} = f (q). \tag{23}$$

The phase portrait of CDS (1) drawn in Fig. 37.1 shows an admissible trajectory which joins the points q_0 and q_1 and passes through the point at infinity of the plane (q^1, q^2).

Controllability of General Bilinear CDS on Plane*

In the present section, by considering the controllability problem for the general bilinear system of second order, we shall demonstrate how the idea of the proposed differential-geometric method (the method of phase portrait) enables us to immediately obtain exhaustive results.

Let the given CDS be a bilinear system

$$\dot{q} = A\,q + u\,B\,q, \quad q \in \{q\} = R^2, \tag{1}$$

where A and B are 2 x 2 square matrices with real entries and u is a scalar control called *admissible* if u = u(t) is a piecewise smooth function with values in R^1. Many physical, physiological and biological processes are described by equations of type (1). Though recent years have seen the publication of a number of works devoted to controllability of linear and bilinear systems (see, for example, [20, 127, 132, 133]), the conditions for controllability of CDS of type (1) in explicit and compact algebraic form were, as far as we know, not obtained.

In what follows, controllability is understood in the usual sense. Set M = $R^2/\{0\}$. System (1) is called *controllable* if the representative point of (1) can attain in a finite time T > 0 a (given) final state q_1 from a given initial state $q_0 \in M$ under the action of the admissible control u = u(t), $0 \le t \le T$.

Before presenting an account of the results obtained, we describe a property of bilinear system (1) which will be referred to as the "symmetry property of hatching" with respect to the origin.

SYMMETRY PROPERTY OF HATCHING. Let q $\in$ M. Then the vector Bq determines the slope of the tangent to the trajectory, passing through the given point q, of the equation

$$\dot{q} = B\,q. \tag{2}$$

* The results of the present section are due to N.L. Lepe.

The trajectories of Eq. (2) in M are boundaries of trajectory funnels of CDS (1). It should be observed that under the action of the admissible control one can move as close to this trajectory as desired on both sides. And when u ---> ∞, we can assume that admissible motion is possible in both directions along trajectory of Eq. (2), that is, on both sides along the boundary of the funnel of CDS (1). The direction of the vector Aq determines hatching of this boundary at q.

We now take a point cq ∈ M, where c ≠ 0 is a numerical constant. The vector cBq is parallel to Bq. Consequently, the tangent to the trajectory of Eq. (2) passing through cq has the same slope at c q as the tangent to the trajectory passing through q. The vectors cAq and Aq are like if c > 0 and unlike if c < 0. This implies that directions of hatching at cq and q coincide if c > 0 and directions are opposite if c < 0. This property is referred to as the *symmetry property* of hatching with respect to the origin of the coordinate system.

We explain the above facts by an example. Let 0 be the origin. The points F, O, C, D are collinear (Fig. 38.1). Through C let a trajectory of Eq. (2) pass on which hatching

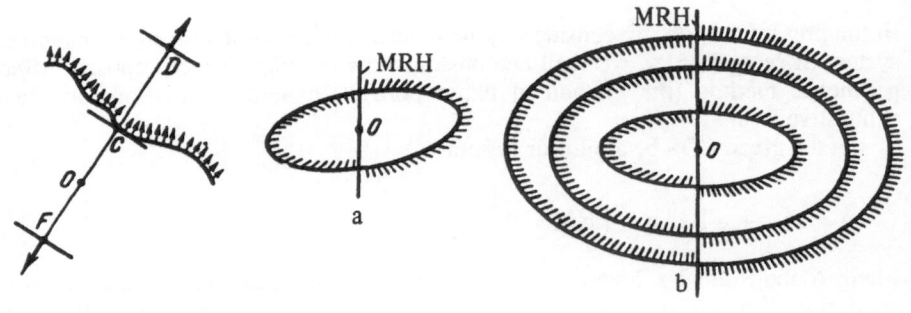

Fig. 38.1   Fig. 38.2

is performed. In this figure the direction of hatching has been shown by arrows. By exploiting the symmetry property of hatching, we can determine, without any computation, the tangents at D and F to the trajectories of Eq. (2) ; they are parallel to the tangent to the trajectory at C. Since the points C and D lie on the segment FD on the same side of O, directions of hatching at these points coincide. The points C and D are situated on opposite sides of O ; therefore directions of hatching at these points are opposite.

Another application of the symmetry property of hatching consists in the following. All properties of a controllable bilinear system can be qualitatively investigated in a spatial domain of finite dimensions and then the results extended to the whole of R^2. For example, suppose we know hatching only on one trajectory of Eq. (2) (Fig. 38.2 a). By exploiting the symmetry property of hatching we can obtain the full picture of hatched trajectory of Eq. (2) in the entire phase plane (Fig. 38.2 b)

We now proceed to investigate the controllability property of CDS (1). This will be done separately for all possible cases of combination of two eigenvalues of the matrix B. Namely, when B has (i) two equal real eigenvalues (ii) two distinct real eigenvalues (iii) two pure imaginary eigenvalues and (iv) two conjugate complex eigenvalues with nonzero real parts. We shall examine these cases one by one.

1. B has two equal real eigenvalues. Keeping aside temporarily the case $B = k\,E$, where k is a scalar and E the identity matrix, we assume that CDS (1) has been transformed to the form

$$\dot{y} = A_1\,y + u\,B_1\,y, \tag{3}$$

where $B_1 = \begin{bmatrix} \lambda & 1 \\ 0 & \lambda \end{bmatrix}$ is the Jordan form of B and λ its repeated real eigenvalue. To discuss the controllability of the given CDS, we examine the family of trajectories of the equation

$$\dot{y} = B_1\,y. \tag{4}$$

These trajectories are the boundaries of the trajectory funnels of CDS (3). The singular point $y = 0$ of this equation is a node. It can be shown that for the bilinear CDS (1) sets of reverse hatching (SRH) in M can be only straight lines, if they exist, not exceeding two in number.

Let us show that CDS (3) can have either two distinct hatching reversal straight line or none. Suppose that CDS (3) has only one hatching reversal straight line. This line is given by the equation

$$|\,A_1 y \quad B_1 y\,| = 0. \tag{5}$$

But, on the other hand, the determinant $|\,[\,A_1\,,\,B_1\,]\,y\,B_1 y\,|$ vanishes on this line implying that the vectors $[\,A_1\,,\,B_1\,]\,y$ and $B_1 y$ are linearly dependent on (5). Hence (5) cannot be a hatching reversal line [127]. This contradiction proves our statement.

Next, assume that CDS (3) has two hatching reversal lines. By the symmetry property its phase portrait will have the form depicted in Fig. 38.3 a to within the direction of hatching. Here the important point to be noted is that there is only one line CD belonging to the family of boundaries of trajectory funnels on which hatching is defined. This line divides the plane into two half-planes, the upper half-plane and the lower half-plane. Such a CDS is, as is clear from the figure, completely controllable,

Indeed, from any trajectory of Eq. (3) lying in the upper half-plane one can move over to any other trajectory also lying in the upper half-plane because every trajectory has a singular point where hatching changes its direction. This statement is also true for any two trajectories (boundaries of funnels) lying in the lower half-plane. Since on CD, a member of the family of trajectories, hatching is defined, one can move from the upper half-plane to the lower half-plane and vice-versa. Hence CDS (3) is controllable. Note that CDS (3) is uncontrollable if hatching is not defined on CD.

Now suppose that (3) has no hatching reversal curve (line). Then its phase portrait is as drawn in Fig. 38.3 b to within the direction of hatching.

In this case also, controllability depends only on whether or not hatching is defined on CD, a member of the family of trajectories of Eq. (3). If hatching is defined, the system is controllable. Indeed, in spite of the fact that from any trajectory in the upper half-plane one cannot move *directly* to any other trajectory, also in the upper half-plane, controllable transition is possible all the same through the lower half-plane. One of such paths is shown in Fig. 38.3 b. Note that if hatching is undefined on CD then CDS is uncontrollable because one cannot go from, say, the upper half-plane to the lower half-plane.

To summarize the above discussion, we can assert that CDS (3) is controllable if hatching is defined on the line CD belonging to the family of trajectories of Eq. (4).

Let us express this controllability condition in an algebraic form. If the matrix B_1 is in the Jordan form, then the straight line belonging to the family of trajectories of Eq. (4) is the line $y^2 = 0$. Hatching on this line cannot be defined if this line belongs to the family of trajectories of the equation $\dot{y} = A_1 y$. But this is possible only if $a_{21} = 0$, where a_{21} is the entry in the second row and first column of A_1

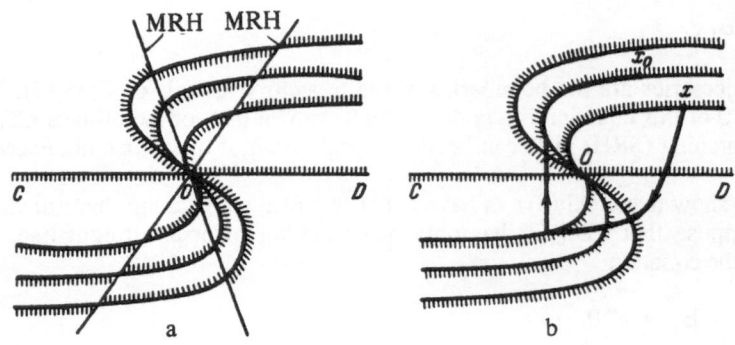

Fig. 38.3

This is the desired necessary and sufficient condition for the controllability of CDS (3) when B is in the Jordan form.

Assume now that B is an arbitrary matrix. Then there exists a nondegenerate linear transformation R such that $B_1 = R B R^{-1}$ is a Jordan matrix. Expressing B_1 in terms of the elements of A and B, we obtain a necessary and sufficient condition for controllability in the form

$$\det [\, A, B\,] < 0. \tag{6}$$

we now take B in the form

$$B = \begin{bmatrix} \lambda & 0 \\ 0 & \lambda \end{bmatrix}.$$

Then trajectories of Eq. (2) constitute a "decrital" node [87].

It should be immediately noted that if the family of trajectories of the equation $\dot{q} = A q$ contains even one straight line, then system (1) is uncontrollable ; hatching is undefined on this line. Since this line divides the phase plane into two disjoint parts, controllable transition of the representative point from one part of the plane to another is not possible. The equation $\dot{q} = A q$ has a trajectory in the form of a line only if the discriminant δ of the

characteristic equation of the given system satisfies the condition
$\delta = (\operatorname{tr} A)^2 - 4 \det \quad A \geq 0$.

We shall show that a necessary and sufficient condition for controllability is of the form

$$\delta = (\operatorname{tr} A)^2 - 4 \det A < 0. \tag{7}$$

Assume that A satisfies (7). Then the singular point of the equation $\dot{q} = A q$ is either a centre or a focus. And every trajectory of $\dot{q} = A q$ intersects all the trajectories of the equation $\dot{q} = B q$.

In moving from the initial point q_0 along a trajectory of $\dot{q} = A q$, the representative point necessarily intersects a trajectory of $\dot{q} = B q$ containing the final point q_1. Further, moving along this trajectory of the equation $\dot{q} = B q$, the representative point reaches the final point q_1. Thus CDS (1) is controllable if condition (7) is satisfied ; otherwise it is uncontrollable, as shown above.

2. Matrix B has distinct real eigenvalues. By a change of variables $y = N q$, where N is a 2 x 2 nondegenerate matrix, CDS (1) is transformed to a CDS of the form

$$\dot{y} \, A_1 y + u \, B_1 y, \tag{8}$$

where $A_1 = NAN^{-1}$, $B_1 = \begin{bmatrix} \lambda_1 & 0 \\ 0 & \lambda_2 \end{bmatrix}$ is a diagonal matrix and λ_1, λ_2 are its distinct real eigenvalues.

The question of controllability of CDS (8) in this case was investigated in [128]. It was proved that CDS (8) is controllable if the elements of A_1 satisfy the condition

$$a_{12} \, a_{21} < 0. \tag{9}$$

The outline of the proof is as follows. In CDS (8) a change of variables

$$\vartheta(t) = \int_0^t u(\tau) \, d\tau, \; z = \exp[-\vartheta B] y. \tag{A}$$

is effected. Then the components z_1 and z_2 of the vector z satisfy the system of equations

$$\dot{z}_1 = a_{11} z_1 + a_{12} z_2 \exp(\Delta \vartheta). \tag{10}$$

$$\dot{z}_2 = a_{21} z_1 \exp(-\Delta \vartheta) + a_{22} z_2,$$

where $\Delta = \lambda_2 - \lambda_1$. Multiplying both sides of the first equation of (10) by z_2 and of the second by z_1 and adding the resulting equations, we obtain

$$\frac{d(z_1 z_2)}{dt} = (a_{11} + a_{22}) z_1 z_2 + a_{12} z_2^2 \exp(\Delta\vartheta) + a_{21} z_1^2 \exp(-\Delta\vartheta). \qquad (B)$$

Solving this equation, we have

$$z_1(t) z_2(t) = \exp(at) \quad [z_1(0) z_2(0) +$$

$$+ a_{12} \int_0^t \exp(-a\tau + \Delta\vartheta) z_2^2(\tau) d\tau +$$

$$+ a_{21} \int_0^t \exp(-a\tau - \Delta\vartheta) z_1^2(\tau) d\tau], \qquad (C)$$

where $a = \operatorname{tr} A_1$. Then the following statements hold : (a) if a_{21} and a_{12} are nonnegative, then $z_1(t) z_2(t) \geq 0$ for $t \geq 0$ and for all $z_1(0) z_2(0) > 0$ independently of the control $v(t)$,

(b) if a_{12} and a_{12} are nonpositive, then $z_1(t) z_2(t) \geq 0$ for $t < 0$ and for all $z_1(0)$ $z_2(0) > 0$ independently of the control $v(t)$.

From (a) and (b) it follows, in particular, that the entire set of points which are accessible from the initial point, not lying on the coordinate axes, coincides with the quadrant containing the initial point. This implies uncontrollability of CDS (1). Somewhat complicated is the proof of controllability of CDS (8) when $a_{12} a_{21} < 0$ [128].

We now demonstrate that (9) can also be obtained by the phase portrait method. For example, when $\lambda_1 \lambda_2 < 0$, the family of trajectories of Eq. (2) has a saddle type singular point. Since every trajectory of this family divides R^2 into two disconnected domains, it is necessary for controllability of CDS (8) that there exists a point on every such trajectory where hatching on this trajectory changes its direction. This means that CDS (8) must have two distinct hatching reversal lines $y^2 = k_1 y^1$ and $y^2 = k_2 y^1$ such that

$$k_1 k_2 < 0. \qquad (11)$$

The equation of hatching reversal curve of CDS (8) is of the form

$$\lambda_2 a_{12} (y^2)^2 + (\lambda_2 a_{11} - \lambda_1 a_{22}) y^1 y^2 - \lambda_1 a_{21} (y^1)^2 = 0. \qquad (D)$$

Since this equation does not contain linear terms in y^1, y^2, hatching reversal curves can be either two distinct lines passing through the origin or one such line. For condition (11) to be satisfied, it is necessary, by Vieta's theorem, that $a_{12} a_{21} < 0$. On the other hand, if CDS (8) has two hatching reversal lines, then the CDS has the form as depicted in Fig. 38.4 to within the direction of hatching. It shows that CDS (8) is completely controllable.

We now raise the following question. What will happen to inequality (9) if B in (8) is not a diagonal matrix ? In order to answer this question, we express the elements of A_1 in terms of those of A and N. Recall that the nondegenerate matrix N is such that $NBN^{-1} = B_1$, where B_1 is a diagonal matrix. It is well known that N, transforming B to the Jordan form, is not unique. But the sign of the product $a_{12} a_{21}$ is independent of the choice of N and is the same as the sign of the expression

$$(a'_{11} - a'_{22}) (b'_{22} - b'_{11}) (a'_{21} b'_{12} + a'_{12} b'_{21}) -$$
$$- (b'_{21} a'_{12} - a'_{21} b'_{12})^2 + a'_{12} a'_{21} (b'_{11} - b'_{22})^2 +$$

$$+b'_{12} b'_{21} (a'_{11} - a'_{22})^2 = \det [A, B],   \tag{A}$$

where $[A, B] = A B - B A$ and a'_{ij}, b'_{ij} are elements of A and B, respectively.

Thus if B has two distinct real eigenvalues system (1) is controllable if

$$\det [A, B] < 0.   \tag{12}$$

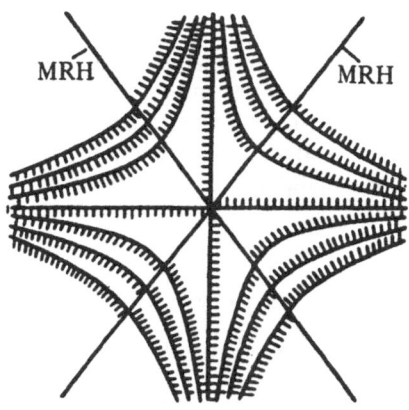

Fig. 38.4

3. Matrix B has two pure imaginary eigenvalues. In this case, the singular point of (2) is a centre. If system (1) does not have hatching reversal curves, then its phase portrait is of the form depicted in Fig. 38.5 a. This figure shows that the CDS is uncontrollable since, for example, we cannot reach any other point of the space from the neighbourhood of the singular point. If, however, (1) has at least one hatching reversal curve, the portrait has the form depicted in Fig. 38.5 b (again to within direction of hatching) in view of the symmetry property of hatching. The system having at least one hatching reversal curve (line) is completely controllable. This is because from any point of M there are controllable paths to a region adjacent to the singular point as well as to any point of R^2 no matter how far this point is from the origin. Therefore the matrices B in question a necessary and sufficient condition for controllability is the existence of at least one hatching reversal curve.

Let us obtain this condition in the analytical from. For the bilinear system, the equation of hatching reversal lines is of the form

$$| Aq \quad Bq | = 0.   \tag{13}$$

Expanding (13) and using the fact that B has pure imaginary eigenvalues, that is, tr B = 0, we obtain necessary and sufficient conditions

$$\det [\, A, B\,] + [\, \text{tr } A\,]^2 \det B < 0. \qquad (14)$$

for controllability. Since the eigenvalues of B are pure imaginary, and hence, det B > 0, the necessary condition for controllability becomes

$$\det [\, A\,, B\,] < O. \qquad (15)$$

4. Matrix B has complex conjugate eigenvalues with nonzero real parts. In this case the singular point of (2) is a focus. We shall show that if hatching is defined at least one point of R^2, then (1) is controllable.

Indeed, suppose that hatching is defined at some point q' ≠ 0. It follows from the symmetry property of hatching that hatching is defined at all points of the line joining the points q' and O (Fig. 38.6). Accordingly, from any point q_0 (not shown in Fig. 38.6), we can travel along a trajectory of Eq. (2) and reach its intersection with the line Oq_0. Since hatching is defined at the point of intersection, we perform controllable passage, in accordance with hatching, to a spiral passing through the final point q_1. Moving further along this spiral one can reach q_1. Thus for controllability of (1) in this case it is necessary and sufficient that hatching be defined at least one point of the space. This controllability condition is equivalent to the condition

$$B \neq kA \qquad (16)$$

for any real k.

Note that if we set $x = (\, a'_{22} - a'_{11}\,)$, $y = (\, b'_{22} - b'_{11}\,)$ then

$$\det [\, A, B\,] = b'_{21} b'_{12} x^2 - (\, a'_{21} b'_{12} + a'_{12} b'_{21}\,) xy +$$

$$+ a'_{12} a'_{21} y^2 - (\, b'_{12} a'_{21} - b'_{21} a'_{12}\,)^2 \qquad (17)$$

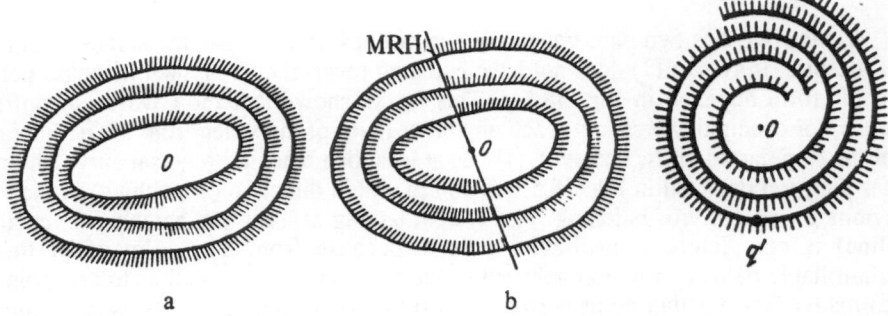

MRH

a   b

Fig. 38.5   Fig. 38.6

We regard (17) as a quadratic expression in x and examine its sign. The discriminant of this expression is transformed to $(\, b'_{12} a'_{21} - b'_{21} a'_{12}\,)^2 (\, y^2 + 4\, b'_{12} b'_{21}\,)$. Since the eigenvalues of B are complex, it follows that $y^2 + 4\, b'_{12} b'_{21} < 0$ and $b'_{12} b'_{21} < 0$. Consequently, (17) is nonpositive. If $B = k\,A$ for some k, then the condition that $\det [\,A, B\,] = 0$ is fulfilled and CDS (1) is uncontrollable. However, it is simple to furnish an example where $\det [\,A, B\,] = 0$ and $B \ne k\,A$. Such a system is controllable. If the strict inequality $\det [\,A, B\,] < 0$ holds, then CDS (1) is controllable.

Thus Paragraphs 1-4 of the present section were devoted to the investigation of controllability conditions for CDS (1) depending on the eigenvalues of B. Relations (6), (7), (12), (14)-(16) yield the following

NECESSARY AND SUFFICIENT CONDITIONS FOR CONTROLLABILITY OF CDS (1).

1. $\det [\,A, B\,] < 0$ if B has real eigenvalues and $B \ne k\,E$ for any real quantity k, where E is the identity matrix.

2. $[\,\mathrm{tr}\,A\,]^2 - 4\det A < 0$ if $B = k\,E$ for some real k.

3. $\det [\,A, B\,] + [\,\mathrm{tr}\,A\,]^2 \det B < 0$ if B has pure imaginary eigenvalues.

4. $B \ne k\,A$ for any real k if B has complex eigenvalues with nonzero real parts.

Finally, a byproduct of our analysis is another valuable result : CDS (1) is uncontrollable if $\det [\,A, B\,] > 0$.

Trajectory Funnel in Backward Time

For the CDS

$$\dot{q} = f(q, u), u \in U(q),$$ (1)

by a *trajectory funnel in backward time* $V_-(q')$ with vertex $q' \in \{q\}$ we mean the funnel $V(q')$ with the same vertex q' drawn for the equation

$$\dot{q} = -f(q, u), u \in U(q).$$ (2)

The idea of constructing transition surfaces in the theory of optimal control by using the "retrograde" motion was advanced by a.A. Fel'dbaum [107, 108, 109].

The associated connection for Eq. (2) is of the form

$$\dot{q} \in -f(q, u).$$ (3)

The cone $K_-(q')$ of admissible velocity directions for (2) is given by the formula $K_-(q') = -K(q')$, where $K(q')$ is the cone of admissible velocity directions for CDS (1). If the Hamiltonian of (1) is $H(p, q)$, then that of (2) is clearly $H(-p, q)$. Note that the trajectory funnel $V(q')$ describes the *domain of readability* of CDS (1) from q', at least for motion from q' during sufficiently small time. In its turn, the trajectory funnel $V_-(q')$ in backward time bounds the *domain of controllability* of the original system (1) for q'.

Sometimes the notion of trajectory funnel in backward time $V_-(q')$ enables us to obtain a constructive solution of the problem of finite control. Suppose that for CDS (1) it is required to find an admissible trajectory from the initial point q_0 to the final point q_1. To solve this problem, we shall construct the boundary of the trajectory funnel $V(q_0)$ for CDS (1) in forward time while for q_1 the boundary of the trajectory funnel $V_-(q_1)$ in backward time. If the two hypersurfaces of boundaries of the funnels $V(q_0)$ and $V_-(q_1)$ intersect at least in one point, then the desired control exists and it can be found in a constructive manner. But is is clear that it will not be unique if $V(q_0)$ and $V_-(q_1)$

intersect in more than one point. However, these two funnels may not directly intersect. Each of them may "press" against domains which can be connected by admissible trajectories. Then the desired control exists. In contrast, when these domains cannot be connected by admissible trajectories, the desired finite control does not exist.

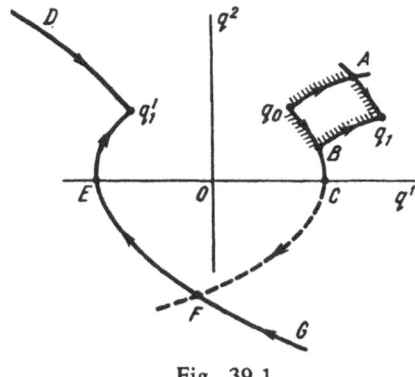

Fig. 39.1.

Note that if the phase portrait of the original CDS (1) has been already drawn, there is no need of additional construction of boundaries of funnels in backward time. They have only to be identified by moving along constructed trajectories passing through the final point q_1 but now in the direction opposite in which t elapses.

As an illustration, we consider the case of the simple CDS

$$\dot{q}^1 = q^2, \quad \dot{q}^2 = u, \quad |u| \le m, \quad m > 0. \tag{A}$$

The phase portrait of this CDS was drawn in Sec. 35. Fig. 39.1 illustrates the idea of solving the finite control problem based on the intersection of forward and backward funnels. For a vivid illustration of this fact two points, the initial point q_0 and the final point q_1, have been taken in the first quadrant. The broken curve A q_0B is the boundary of V (q_0) and the curve Aq_1 B that of the funnel V_- (q_1) in backward time. These two funnels intersect at A and B. Therefore there are at least two admissible trajectories q_0 A q_1 and q_0 B q_1 joining q_0 and q_1.

In another case, when q_0 is in the first quadrant and q_1 is the second, the funnels V (q_0) and V_- (q_1) do not directly intersect. The end point of the boundary of V (q_0) is C and that of ∂V_- (q_1) is E since the q^1 -axis is a singular hatching reversal curve.

Thus in this case the two funnels V (q_0) and V_- (q'_1) "press" against the lower half-plane $q^2 < 0$. But in the lower half-plane there exist, as can be easily seen, admissible trajectories. For example, the trajectory CFE which is not only admissible but also minimal with respect to time.

By constructing the transition curves by the method of "retrograde" motion, due to A.A. Fel'dbaum [109, 110], we can easily find the transition curve for the minimal time problem when we have to move from a given point q $\in$ { q } to a fixed point q'_1. In Fig. 39.1, this transition curve consists of two parabolas Dq'_1 and q'_1 EFG.

40

Optimal Control

Suppose that a CDS is governed by the equation

$$\dot{q} = f(q, u). \tag{1}$$

The CDS (1) is subjected, besides (1), to additional conditions on the values of the control $u \in U(q)$ and on the phase coordinates (or state) $q \overline{\in} G$, where $U(q)$ and G are prescribed sets. The optimal control problem is formulated in the following manner. Given a function $J(q)$ and the initial point $q(O) = q_0$, it is required to find an admissible trajectory for CDS (1) originating from q_0 such that the representative point q assigns the minimum value to $J(q(T))$ at a given instant of time $T > 0$.

Especially simple to solve the problem is in the case of a plane (or a two-dimensional manifold), that is, when the dimension of the state space (manifold) of the CDS is 2. In this case, it suffices to superpose on the phase portrait of the CDS the level surfaces (curves) of the function $J(q)$. Since the phase portrait of the CDS enables us to immediately draw the admissible trajectory, we must choose from the set of admissible trajectories a trajectory along which the representative point q reaches the level curve with the minimum value (possible for this trajectory) of $J(q)$.

The phase portrait of the CDS supplemented by the superposition of level curves of $J(q)$ furnishes a complete qualitative picture of possible solutions of the optimal control problem. Besides, it enables us to obtain the corresponding quantitative characteristics too. The phase portrait drawn on a flat manifold enables us, as demonstrated by Example 3 of the present section, immediately to solve optimal control problems of dimension greater than 2.

We now cite examples illustrating how the suggested approach is realized in solving optimal control problems.[1]

1. For the CDS

---

[1] The examples given below have been suggested and solved by E.A. Andreeva.

$$\dot{q}^1 = -q^1 + u, \quad 0 \le u \le 1, \tag{2}$$

the optimal control problem consists in finding an admissible trajectory from q^1_0 to q^1_1 so that the functional

$$J = \int_0^1 u(t)\,dt \tag{3}$$

attains its minimum value. Introduce a new coordinate

$$\dot{q}^2 = u, \quad 0 \le u \le 1. \tag{4}$$

Then CDS becomes

$$\dot{q}^1 = -q^1 + u, \quad \dot{q}^2 = u, \quad 0 \le u \le 1. \tag{5}$$

For this system it is required to find a trajectory with the initial condition

$$q^1(0) = q^1_0, \qquad q^2(0) = 0, \tag{6}$$

and the final condition, with $T = 1$,

$$q^1(T) = q^1(1) = q^1_1, \tag{7}$$

so that the function $J(q^2)$ attains a minimum for $q^2 = q^2(T) = q^2(1)$.

It can be easily seen that the boundaries of trajectory funnels of CDS (5) are determined by the relations

$$q^2 = C_1, \quad q^2 = C_2 - \ln|1 - q^1|, \tag{8}$$

where C_1 and C_2 are the family parameters. The level curve of $J(q^2_2) \equiv \dot{q}^2$ coincides with the first family of boundaries of trajectory funnels in (8), that is, $q^2 = C_1$. Fig. 40.1 depicts the phase portrait of CDS (5) together with level curves of the functional for which minimum is sought. The desired optimal trajectory joining q_0 and q_1 has been identified in the diagram.

2. For the CDS of the form $\dot{q}^1 = u^1$, $|u^1| \le 1$, with the initial condition $q^1(0) = 0$ and the final condition $q^1(T) = q^1(1) = q^1_1 > 0$, find the optimal trajectory minimizing the functional

$$J = \int_0^1 \left[u^1(t) \right]^2 dt. \tag{A}$$

We introduce the new variable $\dot{q}^2 = (u^1)^2, q^2(0) = 0$, so that the quantity $q^2(T) = q^2(1)$ attains its minimum value.

The boundaries of trajectory funnels are described by the equations $dq^1 / d q^2 = \pm 1$. On integration, we have two families of boundaries, $q^2 = q^1 + C_1, q^2 = -q^1 + C_2$.

Fig. 40.2 depicts the phase portrait of the resulting system. It is clear from the diagram that the desired minimum equals the ordinate of A and the corresponding optimal trajectory is the segment OA.

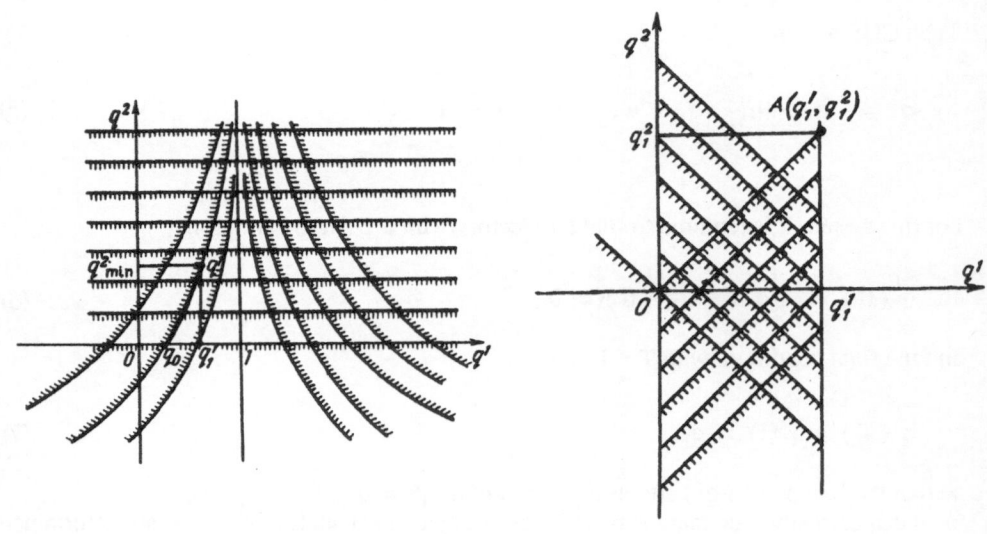

Fig. 40.1   Fig. 40.2

3. We draw a convex figure on the plane which has the minimum (maximum) perimeter subject to prescribed restrictions on its width. We shall describe this figure by means of the notion of the width of a convex figure in a direction φ. The perimeter J is given by the formula

$$J = \int_0^\pi q^1(\varphi) d\varphi. \tag{B}$$

The convexity of the figure is characterized by the condition $q^1 + q^1 = u(\varphi) \geq 0$, where $u(\varphi)$ is the radius of curvature in the direction φ. Constraints on the width $q^1(\varphi)$ are expressed by the conditions $\Delta_o \leq q^1(\varphi) \leq \Delta_1$. If the perimeter of the body is to be minimized, then the boundary conditions are of the form

$$q^1(0) = q^2(\pi) = \Delta_1, \quad q^2(0) = q^2(\pi) = 0, \quad q^2 = \dot{q}^1. \tag{A}$$

If the perimeter is to be maximized, the boundary conditions are $q^1(0) = q^1(\pi) = \Delta_0$.
  Thus we have the following two optimal control problems :

(a) $J = \displaystyle\int_0^\pi q^1(\varphi)\,d\varphi \longrightarrow \inf;$

$$\dot{q}^1 = q^2, \quad q^2 = -q^1 + u, \quad \Delta_0 \leq q^1(\varphi) \leq \Delta_1, \quad 0 \leq \varphi \leq \pi, \tag{B}$$

$$u(\varphi) \geq 0, \quad q^1(0) = q^1(\pi) = \Delta_1, \quad q(0) = q^2(\pi) = 0;$$

(b) $J = \displaystyle\int_0^\pi q^1(\varphi)\,d\varphi \longrightarrow \sup;$

$$\dot{q}^1 = q^2, \quad \dot{q}^2 = -q^1 + u, \quad 0 \leq q^1(\varphi) \leq \Delta_1, \quad 0 \leq \varphi \leq \pi, \tag{A}$$

$$u(\varphi) \geq 0, \quad q^1(0) = q^1(\pi) = \Delta_0$$

Let us find the family of boundaries of trajectory funnels. The equations for the families of funnel boundaries are

$$\frac{dq^1}{dq^2} = -\frac{q^2}{q^1} \quad \text{and} \quad \frac{dq^1}{dq^2} = 0. \tag{B}$$

Integration of these equations yields $q^1 = C_1$, $(q^1)^2 + (q^2)^2 = C_2$, where C_1 and C_2 are the family parameters of boundaries of trajectory funnels. The hatching reversal curve is the q^1-axis because

$$s(q) \equiv |\, Aq\, b\, | = q^2 = 0. \tag{C}$$

Taking into account

$$J = \int_0^\pi q^1 d\varphi = \int_0^\pi (-\dot{q}^2 + u)\,d\varphi = q^2(0) - q^2(\pi) + \int_0^\pi u\, d\varphi \tag{D}$$

and the boundary conditions for problem (a), we find that the optimal phase trajectory starts from the point B (Fig. 40.3), then goes up to C along a circle of radius s_1 with centre O and then goes through the points D and A and once again moving along the circle of radius s_1 with centre O returns to B.

For Problem (b), the optimal phase trajectory on the phase portrait Fig. 40.3 is marked by ABC.

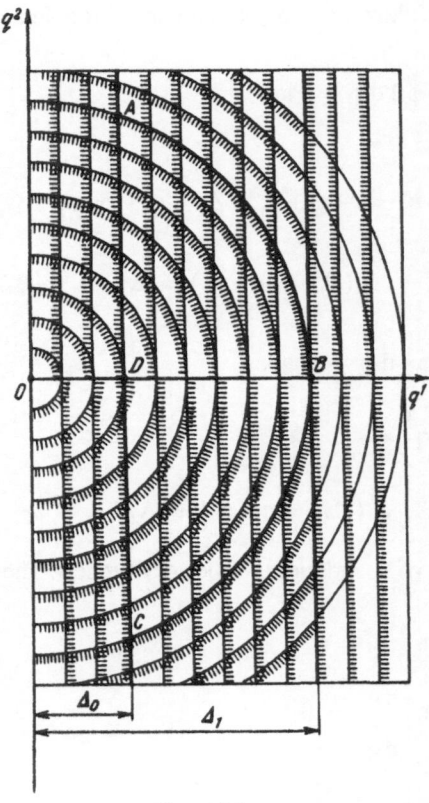

Fig. 40.3

Bibliography

1. M.A. Aizerman   Introduction to dynamics of automatic control of engines, Izd. Mashgiz, Moscow 1950.
2. M.A. Aizerman and E.S. Pyatnitskii,   Lectures on analytical mechanics, Nauka, Moscow 1981.
3. P.S. Aleksandrov   Introduction to general theory of sets and functions, Izd. Gostekhteorizdat, Moscow 1948.
4. Yu. N. Andreev   Control of finite-dimensional linear objects, Nauka, Moscow 1976.
5. A.A. Andronov, A.A. Vitt and E.S. Haikin   Theory of vibrations, Fizmatgiz, Moscow 1959.
6. A.A. Andronov E.A. Leontovich, I.I. Gordon and A.G. Maier   Qualitative theory of second-order dynamical systems, Nauka, Moscow 1966.
7. A.A. Andronov, E.A. Leontovich, I.I. Gordon and A.G. Maier   Theory of bifurcations of dynamical systems on plane, Nauka, Moscow 1967.
8. V.I. Arnol'd   Supplementary chapters on differential equations, Nauka, Moscow 1972.
9. V.I. Arnol'd   Mathematical methods of classical mechanics, Nauka, Moscow 1974.
10. V.I. Arnol'd   Special features of differential mappings, Nauka, Moscow 1982.
11. R. Bellman and E. Engel   Dynamic programming and partial differential equations, Mir, Moscow 1975.
12. V.V. Belov and E.M. Vorob'ev   Collection of problems on supplementary chapters of mathematical physics, Izd. Vysshaya Shkola, Moscow 1978.
13. G. Birkhoff   Dynamic systems, Gostekhteorizdat, Moscow-Leningrad 1941.
14. V.I. Blagodatskikh   Theory of differentiable connections, Vol. I, Izd. Mos. Gosu. Univ., Moscow 1979.
15. D.I. Blokhintsev   Space and time in microworld, Nauka, Moscow 1982.
16. N.N Bogolyubov and Yu. A. Mitropol, skii   Asymptotic methods in theory of nonlinear vibrations, Nauka, Moscow 1974.
17. B.M. Bolotovskii and S.I. Stolyarov   Fields of sources of illumination in moving media, in Einstein collection of articles, 1978-1979, pp. 197-277, Nauka, Moscow 1983.
18. V.G. Boltyanskii   Optimization problem with change in phase space, Differentsial'nye Uravneniya, 19 (March 1983), 518-520.

18a.	V.G. Boltyanskii	Method of local sections and principle of support, in the book "Mathematics in service of engineer", Izd. Znanie, Moscow 1973.
19.	M. Born and E.Wolf	Foundations of optics, Nauka, Moscow 1970.
20.	P.W. Brockett	Lie algebras and Lie groups in theory of control, in collection "Mathematical methods in system theory", Mir, Moscow 1979.
21.	G. Buzeman	Convex hypersurfaces, Nauka, Moscow 1964.
22.	A.G. Butkovskiy	Theory of optimal control of systems with distributed parameters, Nauka, Moscow 1965.
22a.	A.G. Butkovskiy	Distributed control systems, Elsevier Publishing Co., New York, 1969 (English version of [22]).
23.	A.G. Butkovskiy	Methods of control of systems with distributed parameters, Nauka, Moscow 1975.
24.	A.G. Butkovskiy	Structure theory of distributed systems, Nauka, Moscow 1977.
24a.	A.G. Butkovskiy	Structural theory of distributed systems, Ellis Horwood Publishers, Chichester, England, 1983 (English version of [24]).
25.	A.G. Butkovskiy	Characteristics of systems with distributed parameters, Nauka, Moscow 1979.
25a.	A.G. Butkovskiy	Green's functions and transfer functions (Hand Book), Ellis Horwood Publishers, Chichester, England, 1982 (English version of [25]).
26.	A.G. Butkovskiy	Differential-geometric method of constructive solution of problems of controllability and finite control, Avtomatika i Telemekhanika 1 (1982).
27.	A.G. Butkovskiy, Yu.N. Andreev and S.A. Malyi	Optimal control of metal heating, Izd. Metallurgiya, Moscow 1972.
28.	A.G. Butkovskiy, Yu.N. Andreev and S.A. Malyi	Control of metal heating, Izd. Metallurgiya, Moscow 1981.
29.	A.G. Butkovskiy, and L.M. Pustyl'nikov	Theory of mobile control of systems with distributed parameters, Nauka, Moscow 1980.
29a.	A.G. Butkovskiy, and L.M. Pustyl'nikov	Mobile control of distributed parameter systems, Ellis Horwood Publishers, Chichester, England, 1987 (English version of [29]).
29b.	A.G. Butkovskiy, and Yu.I. Samoilenko	Control of quantum processes, to appear, Kluwer Academic Publishers, Netherlands, 1990 (English version of [30]).
30.	A.G. Butkovskiy, and Yu.I. Samoilenko	Control of quantum mechanical processes, Nauka, Moscow 1983.
30a.	J.Wolf	Constant curvature spaces, Mir, Moscow 1982.

31.	A.A. Voronov	Foundations of theory of automatic control, Izd. Energiya, Moscow-Leningrad, 1966.
32.	A.A. Voronov	Stability, controllability and observability, Nauka, Moscow 1979.
33.	R. Gabasov and F.M. Kirillova	Qualitative theory of optimal processes, Nauka, Moscow 1971.
34.	R. Gabasov and F.M. Kirillova	Special optimal controls, Nauka, Moscow 1973.
35.	I.M. Gel' fand and S.V. Fomin	Calculus of variations, Nauka, 1961.
36.	S.K. Godunov	Equations of mathematical physics, Nauka, Moscow 1974.
37.	V.I. Gurman	Degenerate problems of optimal control, Nauka, Moscow 1977.
38.	E. Goursat	Course of mathematical analysis, Vols. I, II, Gostekh-teorizdat, Moscow-Leningrad 1936.
39.	N.M. Gyunter	Integration of first-order partial differential equations, Gostekhteorizdat, Moscow-Leningrad 1934.
40.	Yu.L. Daletskii and S.V. Fomin	Measures and differential equations in infinite-dimensional
41.	P. Dirac	Generalized Hamiltonian mechanics, in the collection "Variational principles of mechanics" (Editor : L.S. Polak), pp 705-722, Fizmatgiz, Moscow 1959.
42.	P. Dirac	Principles of quantum mechanics, Nauka, Moscow 1972.
43.	B.A. Dubrovin, S.P. Noviknv and A.T. Fomenko	Modern geometry, Nauka, Moscow 1979.
44.	G. Dyuvo and J. Lions	Inequalities in mechanics and physics, Nauka, Moscow 1980.
45.	S.V. Emel'yanov	Automatic control systems with variable structure, Nauka, Moscow 1967.
46.	N.V. Efimov	Introduction to theory of exterior forms, Nauka, Moscow 1977.
46a.	A.F. Filippov	Differential equations with discontinuous right side, Nauka, Moscow, 1985
47.	A.A. Zhevnin and A.P. Krishchenko	Controllability of nonlinear systems and synthesis of of control algorithms, Dokl. Akad, Nauk SSSR 258 (1981), No.4.
48.	V.P. Zhukov	Investigation of stability of a class of nonlinear systems, Avtomatika i Telemekhanika 11 (1980).
49.	L. Zade and C. Desoer	Theory of linear systems, Nauka, Moscow 1978.
50.	V.A. Zagaller	Theory of envelopes, Nauka, Moscow 1975.
51.	E. Kamke	Handbook of first-order partial differential equations, Nauka, Moscow 1966. (German original : Leipzig 1959).
52.	G. Carslaw and D. Eger	Heat conduction in solid bodies, Nauka, Moscow 1964.

53.   E. Cartan   Integrable invariants, Gostekhteorizdat, Moscow-Leningrad
                             1940.
54.   A.M. Kovalev   Nonlinear control and observation problems in theory of
                             dynamical systems, Naukova Dumka, Kiev 1980.
55.   M.A. Krasnosel'skii,   Vector fields on plane, Fizmatizdat, Moscow 1963.
56.   M.A. Krasnosel'skii,   Topological methods in theory of nonlinear equations,
                             Gostekhteorizdat, Moscow 1956.
57.   M.A. Krasnosel'skii   Geometric methods in nonlinear analysis, Nauka, Moscow
      P.P. Zabreiko,   1975.
58.   M.A. Krasnosel'skii,   Vector fields on plane, Nauka, Moscow 1961.
      A.I. Perov,
      A.I. Povolotskii and
      P.P. Zabreiko,
59.   M.A. Krasnosel'skii   Convex functions and Orlicz spaces, Fizmatgiz, Moscow
      and   1958.
      Ya. B. Rutitskii,
60.   N.N. Krasovskii,   Theory of control of motion, Nauka, Moscow 1968.
61.   N.N. Krasovskii and   Positional differentiable games, Nauka, Moscow 1974.
      A.I. Subbotin,
62.   R. Courant,   Partial differential equations, Mir, Moscow 1964.
63.   R. Courant and   Methods of mathematical physics, Vols. I,II,
      D. Hilbert,   Gostekhteorizdat, Moscow-Leningrad 1945. (English
                             edition : Interscience, New York, 1953, 1962)
63a. M.A. Lavrent'ev and   A course of calculus of variations, Gostekhteorizdat,
      L.A. Lyusternik,   Moscow 1950.
64.   Yu.P. Ladikov,   Stabilization of processes in continuous media, Nauka,
                             Mosocw 1978.
65.   L.D. Landau and   Mechanics, Nauka, Moscow 1965.
      E.M. Lifshits
66.   T. Levi-Civita and   A course of theoretical mechanics, Vol. II, Izd. Inost. Lit.,
      W. Amaldi,   Moscow 1951.
67.   V.S. Levchenknov   On general theory of system, Izd. Vsesoyuz. Naucno-Issled.
      and A.I. Propoi   Instituta Systemnikh Issledovanii (VNIISI), Moscow 1978.
68.   J. Leightman   Introduction to theory of optimal control, Nauka, Moscow
                             1968.
69.   M.A. Leontovich   Statistical physics, Gostekhizdat, Moscow 1944.
70.   J. Lere   Lagrangian analysis and quantum mechanics, Mir, Moscow
                             1981.
71.   J.W. Lich   Classical mechanics, Izd. Inost. Lit., Moscow 1961.
72.   K.A. Luri'e   Optimal control in problems of mathematical physics,
                             Nauka, Moscow 1975.
73.   L.I. Mandel'shtam   Lectures on vibrations, Izd. Akad. Nauk SSSR, Moscow
                             1955.
74.   V.P. Maslov   Operational methods, Nauka, Moscow 1976.
75.   A.S. Mishchenko,   Lagrangian manifolds and method of canonical operator,
      B.Yu. Sternin and   Nauka, Moscow 1978.
      V.E. Shatalov

76.	E.F. Mishchenko	Asymptotic computation of periodic solutions of systems of differential equations, Izv. Akad. Nauk SSSR (Ser : mat.) 21 (1957).
77.	N.N. Moiseev	Asymptotic methods in nonlinear mechanics, Nauka, Moscow 1981.
78.	Yu.I. Neimark and N.A. Fufaev	Dynamics of nonholonomic systems, Nauka, Moscow 1967.
79.	Yu.I. Neimark and N.A. Fufaev	Nonlinear waves (self organised), Collection of articles, Nauka, Moscow 1983.
80.	V.V. Nemytskii and VV. Stepanov	Qualitative theory of differential equations, Gostekhteorizdat, Moscow-Leningrad 1947. English transl; Princeton 1960).
81.	A.V. Netushil	Theory of automatic control, Izd. Vysshaya Shkola, Moscow 1983.
82.	Yu.L.Pavlovskii and G.N. Yakovenko	Groups admitted by dynamic systems, in the collection "Optimization methods and their applications", Nauka, Novosibirisk 1982.
83.	I.G. Petrovskii	Lectures on ordinary differential equations, Gostekhteorizdat, Moscow-Leningrad 1952.
84.	I.G. Petrovskii	Lectures on partial differential equations, Fizmatgiz, Moscow 1961.
85.	E.S. Polovinkin	Elements of theory of multiple-valued mappings, Izd. Mosk. Fiziko-Tekhn. Instituta, Moscow 1982.
86.	L.S. Pontryagin	On dynamic systems that are close to Hamiltonian systems, Zh. Eksper. i Teori. Fizika, 4 (1934).
87.	L.S. Pontryagin	Ordinary differential equations, Nauka, Moscow 1982.
88.	L.S. Pontryagin, Boltyanskii R.V. Gamkrelidze and E.F. Mishchenko	Mathematical theory of optimal processes, V.G. Fizmatgiz, Moscow 1961. (English transl; Wiley, New York 1962).
89.	E.P. Popov	Dynamics of automatic control system, Gostekhteorizdat, Moscow 1954.
90.	W. Porter	Modern foundations of general theory of systems, Nauka, Moscow 1971.
90a.	M.M. Postnikov	Lectures on geometry. Lie groups and algebras, Nauka, Moscow 1982.
91.	A. Poincare,	On curves defined by differential equations, Gostekhteorizdat, Moscow-Leningrad 1947.
92.	V.S. Pugachaev, I.E. Kazakov and L.G. Evlanov	Foundations of statistical theory of automatic systems, Izd. Mashinostroenie, Moscow 1974.
93.	B.S. Razumikhin	Physical models and methods of equilibrium theory in programming and economics, Nauka, Moscow 1975.
94.	B.S. Razumikhin	Problems of analytical mechanics and optimal control problem, I,II, Avtomatika i Telemekhanika 2, 3 (1976).
95.	P.K. Rashevskii	Geometric theory of partial differential equations, Gostekhteorizdat, Moscow-Leningrad 1947.

96.	L.I. Rozonoer and A.M. Tsirlin	Optimal control of thermodynamical processes, I-III, Avtomatika i Telemekhanika 1, 2, 3 (1983).
97.	R.T. Rockafellar	Convex sets, Mir, Moscow 1973.
98.	H. Rund	Differential geometry of Finsler spaces, Nauka, Moscow 1981.
99.	Yu.I. Samoilenko and V.L. Volkovich	Spatially distributed adoptive and control systems, Tekhnika, Kiev 1968.
100.	J.A. Synge	Classical mechanics, Fizmatgiz, Moscow 1963.
101.	V.I. Smirnov	A course of higher mathematics, Vol. IV, Gostekhteorizdat, Moscow-Leningrad 1951.
102.	V.V. Stepanov	A course of differential equations, Gostekhteorizdat, Moscow 1953.
103.	M.A. Sultanov	Investigation of control processes described by equations having indefinite functional parameters, Avtomatika i. Telemekhanika, 10(1980).
104.	F. Tricomi	Lectures on partial differential equations, Izd. Inost. Lit., Moscow 1957.
105.	V.I. Utkin	Zero overshoot responses and their applications to systems with variable structure, Nauka, Moscow 1974.
106.	J. Javard	A course of local differential geometry, Izd. Inos. Lit., Moscow 1960.
107.	A.A.Fel'dbaum	Electrical systems of automatic control, Izd. Oborongiz, Moscow 1957.
108.	A.A. Fel'dbaum	Computers in automatic systems, Fizmatgiz, Moscow 1959.
109.	A.A. Fel'dbaum	Foundations of theory of automatic systems, Second edition, Nauka Moscow 1966.
110.	A.A. Fel'dbaum and A.G. Butkovskiy	Methods of theory of automatic control, Nauka, Moscow 1971.
111.	A.F. Fillipov	On some questions of the theory of optimal control, Vestnik Mosk. Gos. Univ. (ser:Mat.) 2(1959).
112.	A.F. Fillipov	Differential equations with discontinuous right side, Math. Sb., 51 (1968), 99-108.
113.	L.N. Fitsner	Coordinate control of motion, Nauka, Moscow 1971.
114.	L.N. Fitsner	Biological search systems, Nauka, Moscow 1977.
115.	V.A. Fok	Theory of space, time and gravity, Fizmatgiz, Moscow 1961.
116.	Ya.Z. Tsipkin	Foundations of theory of automatic systems, Nauka, Moscow 1977.
117.	F.L. Chernous' ko	Guaranteed optimal estimates of uncertainties by means by ellipsoids, Tekh. Kibernetka 6(1980).
118.	F.L. Chernous'ko	Ellipsoidal estimates of accessibility domain of controllable system, Prikl. Mat. i. Mekh., 45(1981), No.1.
119.	A.E.El' sgol'ts	Differential equations, Gostekhteorizdat, Moscow 1957.
120.	O.Young	Lectures of calculus of variations and theory of optimal control, Mir, Moscow 1974.

121.	M. Ash	Nuclear reactor kinetics, New York, McGraw-Hill, 1965.
121a.	H. Bresis	Operations maximaux monotones et semi-groupes de contraction dans les espaces de Hilbert, Amsterdam, North-Holland, 1973.
122.	A.G. Butkovskiy	Distributed control systems, New York, Elsevier, 1969.
123.	A.G. Butkovskiy	Green's functions and transfer functions, Chichester, Ellis Horwood, 1982.
124.	A.G. Butkovskiy	Some new results in distributed control systems (survey), 3rd International symposium on distributed parameter systems, Toulouse (Frans.), 1982.
125.	A.G. Butkovskiy	Structural theory of distributed systems, Chichester, Ellis Horwood, 1983.
125a.	A.G. Butkovskiy and L.M. Pustylnikov	The mobile control of distributed parameter systems, Chichester, Ellis Horwood, 1983.
126.	R.F. Curtain and A.J. Pritchard	Infinite dimensional linear systems theory, Berlin, Heidelberg, New York, Springer-Verlag, 1978, 298 pp.
127.	L.R. Hunt	Global controllability of nonlinear systems in two dimensions, Math. Systems Theory, 1980, 13, p. 361–376.
128.	V. Jurdjevic and J.P. Quinn	Controllability and stability, J. Differential Equations, 1978, 28, p. 381–389.
129.	J.-L. Lions	Sur les systemes distributes singuliers, Los Alamos, 1981.
130.	J.-L. Lions	Some aspects of the optimal control of distributed parameter systems, Philadelphia, USA, 1981.
131.	J.-L. Lions	Some methods in the mathematical analysis of systems and their control, INRIA, France, 1981.
132.	R.E. Rink and R.R. Mohler	Completely controllable bilinear systems, SIAM J. Control, 1968, vol. 6, No. 3.
133.	H. Sussman and V. Jurdjevic	Controllability of nonlinear systems, J. Differential Equations, 1972, 12, p. 95–116.

Index